The Trouble with Paradise

The Trouble with Paradise

A HUMOROUS ENQUIRY INTO THE PUZZLING
HUMAN CONDITION IN THE 21ST CENTURY

Dr Robin Lincoln Wood

AuthorHouse™ UK Ltd.
1663 Liberty Drive
Bloomington, IN 47403 USA
www.authorhouse.co.uk
Phone: 0800.197.4150

© 2014 Dr Robin Lincoln Wood. All rights reserved.

No part of this book may be reproduced, stored in a retrieval system, or transmitted by any means without the written permission of the author.

Published by AuthorHouse   07/01/2014

ISBN: 978-1-4969-7506-5 (sc)
ISBN: 978-1-4969-7505-8 (hc)
ISBN: 978-1-4969-7507-2 (e)

Any people depicted in stock imagery provided by Thinkstock are models, and such images are being used for illustrative purposes only. Certain stock imagery © Thinkstock.

This book is printed on acid-free paper.

Because of the dynamic nature of the Internet, any web addresses or links contained in this book may have changed since publication and may no longer be valid. The views expressed in this work are solely those of the author and do not necessarily reflect the views of the publisher, and the publisher hereby disclaims any responsibility for them.

Table of Contents

Foreword xi

Introduction xv

Chapter One
Why Some Like it Hot and Opposites Attract 1

Chapter Two
Great Questions, Big Thinkers, Many Answers 15

Chapter Three
A Good Life, in the 21st Century? 64

Chapter Four
The Trajectory of our Species 103

Chapter Five
Paradise Began in a Garden 133

Chapter Six
On the Origins of Heaven and Hell 143

Chapter Seven
The Arc of Human Development in the 21st Century 190

Chapter Eight
New Answers to Age Old Mysteries 215

Chapter Nine
Overview of a Transpersonal Learning Journey 233

Epilogue: Part One 291

Epilogue: Part Two
The Story of ThriveAbility so Far.. 294

Endnotes.. 341

Detailing some of the recent trouble in Paradise, whereby the author discovers several varieties of Hell, while attempting to create his own small corner of Heaven on Earth.

Along with an explanation of why evolution hurts mainly when we laugh, and how we might mature as a species so that our emerging global civilization can thrive well into the future.

And also, a brief explanation as to why though things have never looked so dark for our biosphere and future generations, we will make it through to living in a renewable-energy-powered, joyful civilization.

Intertwined with an enquiry into why we are so stressed out and frequently unhappy when we have "never had it so good". And why we struggle to find meaning and purpose in life, while being better informed and educated than any generation, ever.

And not to forget the most daring adventure of them all, the journey to the interior of the multiverse, finding new answers to age old mysteries.

And, what one might do about that, exactly . . . .

Heaven is not a location but refers to the inner realm of consciousness. A new heaven is the emergence of a transformed state of human consciousness, and 'a new earth' is its reflection in the physical realm.

Eckhart Tolle

Processes of evolution, including variation and natural selection, niche creation and co-evolution, even catastrophe and fluctuating rates of evolutionary change, suggest that adaptation is usually imperfect, with abundant glitches that, as long as they don't constitute abject failures, usually continue to exist unless selection and variation conspire to find a way to get rid of them.

Neuroanthropologist Greg Downey

Have you ever asked yourself one or more of these questions about life?

1. Who am I, and why am I here? How can I make sense of all the conflicting religious, philosophical and self-help advice people seem so happy to dispense?

2. Am I more than mere chemistry conspiring to be human? Are there other dimensions in the universe beyond my five senses?

3. What is the meaning of life? How can I realize my full potential while I'm here?

4. Why is there so much suffering and conflict on our planet? Why does God/the Universe/whatever allow this to happen?

5. What is the "good" in all our newfound scientific, technological and material progress and power? If we have never had it "so good", then why are we all so stressed out?

6. Are we doomed by climate change and our negative environmental impacts?

7. What does it mean to live a good life? How can I find my "Bliss"?

8. What might a thriving world look like? What can I do to make the world a better place?

Within these pages you will find a clear yet humorous investigation into each of these big questions, which will take you at least halfway to coming up with your own answers to help you make your life more of a joyful adventure. Enjoy!

Foreword
by Ken Wilber

Robin Wood's *The Trouble with Paradise* points out that, among other things, each succeeding unfolding of evolution involves the creation of a greater complexity that is held together by a greater wholeness. Quarks go to atoms, atoms go to molecules, molecules to cells, cells to organisms, and so on—each more complex or sophisticated, each more unified or whole (since each actually contains the previous level of complexity as an actual ingredient in its own higher complexity). Robin's own approach to life—its questions, challenges, issues, problems—follows the same evolutionary course: look for more sophisticated, more inclusive, more "wholistic" approaches, approaches that don't limit themselves to single, narrow, partial, and ultimately fragmented views, but ones that include the "true but partial" aspects of as many relevant areas as possible, thus helping ensure the greatest likelihood of successes, solutions, resolutions.

Jean Gebser called such an approach "integral-aperspectival," a bit of a mouthful but actually easy enough to understand. Gebser was a brilliant pioneer in following the evolution and development of human cultures, and he maintained that humanity was actually on the verge of a new and profound evolutionary leap, from a mental-rational view of reality to the above-mentioned "integral-aperspectival"—a leap that Gebser (among several others, it turns out) felt was perhaps the most monumental leap in humankind's entire history. The "aperspectival" part means "without perspective," or more accurately, utilizing as many perspectives as possible but without privileging any of them. The numerous perspectives ensure that we take into account as many viewpoints of the situation as possible, thus ensuring we are more inclusive and comprehensive in our coverage of all the various aspects of reality that are actually involved (thus, again, increasing our actual likelihood of success). But these perspectives aren't just thrown together in an eclectic or randomly pluralistic fashion; they are all held together in a broader wholeness by an "integrating" or "integral" framework or attitude—again, the very nature of evolution itself. But what was different about this coming stage of evolution for humanity was that this "wholeness" was actually becoming conscious, becoming something that humans were directly aware of, which would increase their own sense of wholeness, unity, fullness, self-actualization, fulfilment.

One of those "others" who happened to agree with Gebser was the pioneering developmental genius Clare Graves, who defined the coming wholeness stage in essentially the same way Gebser did, and added that it was "a monumental leap of meaning, an unbelievable chasm of meaning . . . ." Abe Maslow was another who saw it coming; he called the shift the great leap from "deficiency needs" to "Being needs"—the leap from operating always as if one doesn't have enough of anything to operating from the profound sense that one is overflowing, is drenched in superabundance, in radical fullness, not depressive emptiness and scarcity. Evolution, in becoming conscious of itself (which it just has), is also set to become conscious of its own operations, including the successive unfolding of higher and higher wholes . . . . This won't descend on humanity in an all-at-once avalanche; rather, humans will still have to grow and develop through their earlier stages of growth, on the way to this, the latest stage, an "integral-aperspectival" stage, a "wholistic" unfolding, a monumental leap in meaning . . . .

That is the edge from which Robin writes. And thus every area he steps into, as you will see in the following pages, he outlines an "integral-aperspectival" approach, which, in many cases, happens to include believable solutions to some very serious problems, simply because all of the best approaches are being included in the overall vision—something that, prior to this stage in evolution, anyway, was an extremely rare idea (and is still extremely rare, which is as unfortunate as it is actually dangerous: broken and fragmented approaches give broken and fragmented solutions. Prime evidence: today's world, alas.) But it doesn't have to be that way . . . .

So you'll learn about "eight primordial perspectives," or the most fundamental perspectives or lens that all things can be looked at through, since they are basic ingredients of the universe itself (including things like the Good, the True, and the Beautiful—which are each real dimensions of reality because each is the result of looking at the same reality through a different lens or perspective). Leave any of those out, and you've got a partial, torn, fragmented result (no wonder those approaches never reach any real solutions).

And did you know that, if you take all of the major developmental psychology models ever proposed, they all point to essentially the same 8-10 levels of development? This means in you, in me, in a Chinese person, a Mexican. That's an important piece of information, wouldn't you say? It's a road map that applies to all human beings, research consistently shows, and thus a road map that can not only help us understand where we are now, but where we are headed. Fairly important stuff, yes?

Humans also, as we continue to look at all perspectives, don't have just one major intelligence (generally called "cognitive" and measured with the all-important IQ). But, as Howard Gardner has made popular, humans have "multiple intelligences," including, as Robin summarizes them, a bodily or physiological intelligence, an intuitive intelligence, a rational intelligence, and a spiritual intelligence. And those intelligences all grow and develop through those major 8-10 levels of development, so that you can be highly developed in some intelligences, mediumly developed in others, and poorly developed in yet others. But, as you'll also see, the important item is that there are exercises and practices available that can help you accelerate your growth and development in any intelligence through any of the levels. Hot stuff indeed!

And so on as Robin explores the many various areas he investigates. Which are, what, exactly? Well, sort of everything. As the subtitle suggests, "A humorous enquiry into the puzzling human condition in the 21st century." What you can be assured of is that, in every area he explores with you, he gives a broad and inclusive overview of the various perspectives available on that area—an "integral-aperspectival" approach to each. This includes, for example, the recent expansion of science to include not only the eye of flesh and the eye of mind, but the eye of spirit. Why is there a science in each? Because each is capable of delivering direct, immediate "data" or direct experiences when exercised correctly—a sensory experience, a mental experience, a spiritual experience, each of which is capable of being checked and investigated by others using the same eye, thus producing a verification or rejection of the data—science indeed. The battle between science and spirituality, which was always fairly ridiculous, is finally rendered obsolete.

And what is the "trouble with paradise"? As Robin summarizes it, it's the very nature of evolution to continually be unfolding to higher and more inclusive views, and thus the present view, whatever it is and at whatever level, is always "true but partial"—there's always a "true," or paradise, and always a "partial," or trouble. And the only cure to that is to get with the program itself, accepting this "transcend and include" as the very nature of evolution itself, and thus to begin moving in exactly the same direction, hence riding from paradise to paradise while rejecting trouble after trouble—a grand adventure of continuous wholistic wave surfing.

If human beings are going to actually make it through the 21st century—and not just surviving, but "thriving"—then an integral-aperspectival approach is certainly starting to appear to be the only truly viable path forward. Robin has created "Renaissance2.eu" to explore just such an option, and all of you are invited to drop in and share these notions with him. In the meantime, take the following pages as in introduction to this "monumental leap in meaning" that does, indeed, appear to be the face of the future for all of us.

Ken Wilber
The Integral Vision
Denver CO
Summer 2014

INTRODUCTION

> "Gloriously situated by life at this critical point in the evolution of humankind, what ought we to do? We hold earth's future in our hands. What shall we decide?"
>
> *Pierre Teilhard de Chardin*

I have written, and read, my fair share of very serious, worthy books. This is not one of them. Worthy, yes, but where is the joy in high seriousness? This book is meant to be fun, useful and highly stimulating. **Fun**, because we should never take ourselves, nor life itself, too seriously. **Useful**, because there is nothing so practical as a good philosophy and new ideas suited to the times. **Stimulating**, because the world is a marvelous place full of amazing phenomena, people, places and ideas, and it would be a pity to read a book that did not inspire us to look around us with fresh eyes and new perspectives, and to then do the right things as a result.

As Teilhard de Chardin's quote above says, we are living at a critical juncture in the evolution of our species. We hold the future in our hands in a very special way that is unique to the generations alive in the 21st century. Why? Because those alive in the 21st century will determine, through their thoughts and actions, whether or not our species is pushed to the edge of extinction (along with millions of other species), or whether we will create a thriving future for our descendants and the world at large.

There, that is the serious bit out of the way. One of the key points I make in this book is that we are more likely to be successful at this very serious task, if we do not take ourselves too seriously. We are going to need our sense of humour, creativity and bouncy confidence more than ever before. We are going to need to be able to inspire others to give of their very best, and then more, if we are to survive and thrive into the 22nd century and beyond.

Our species has existed in its current form for little more than 50 000 years, a very short time in the Big History book in which life began some 3.5 billion years ago. To put that into perspective, the dinosaurs roamed the earth as the top predators for over 135 million years. These "terrible lizards" (though they are actually reptiles with elevated body temperatures and the capacity for social interaction), were, in evolutionary terms, highly successful.

The fact that more than 99.99% of all species, languages, nations and organizations that have ever existed, are now extinct, should give us pause for thought. The dinosaurs did not possess a neocortex as we do, but for those 135 million years that they played top predator, things were less complex than they are today, so their reptilian brain stem and primitive limbic systems were more than enough to ensure their evolutionary success until the meteor hit. Today evolution is playing out in the human neocortex and frontal cortex in particular, through our consciousness, culture, socioeconomic systems and technologies. Will climate change be the "meteor" that "hits" us?

The idea of paradise embodies a very ancient idea—finding a place where we can meet our all our needs and enjoy a rich, peaceful and fulfilling existence together. Five centuries ago, when most of life was still nasty, brutish and short, paradise was considered to be possible mainly after death, in a place known as "heaven". The thought of going to Heaven provided a strong incentive to most religious believers, even the most cynical, to follow many of the teachings of their particular religion. Otherwise the yawning, red hot gates of Hell (remarkably similar in many respects to the worst aspects of daily suffering in those times) beckoned.

The famed German philosopher Nietzsche pronounced in 1882 that "God is Dead" (reprised by Time magazine on its famous cover in 1966: "Is God Dead?"). What did he mean, exactly? Nietzsche was actually concerned that the decline of religion might mean a rise in nihilism and anarchy, even though he personally did not believe in God. Au contraire, French existentialist philosopher Albert Camus thought it absurd that human beings craved a higher order.

While such debates have raged hot and furious over the centuries, there is no doubt that the comfortable, secure life style of most of us are now living in the developed world has led to us consider it possible for us to experience heaven on earth, now, rather than having to wait until we die. At least that is what the advertisements imply: You can have it all, <u>now</u>. After all, you're worth it, darling. We need no longer attend church, synagogue, or the mosque to have a shot at heaven.

While numbers at Buddhist sanghas continue to rise worldwide, the Buddhists do not believe in God—kindness, as the Dalai Lama puts it, is their religion.

Yet, as the title of this book suggests, there is considerable Trouble with Paradise. While we might each fantasize about our ideal world, partner, career, life and so on, and even be mightily encouraged in this fantasy by those seeking our support, custom and cash, things do not always work out that way. We find ourselves having to deal with a "Reality" very far removed from the world of our fantasies. And there is also much trouble, even in the cozy, high altitude worlds of sanghas and integral philosophy whose very aim is to transcend and include the trouble with paradise . . . .

The central theme of this book is: "Evolution has ever been thus. Imperfect adaptation drives the evolution of ever more complex, conscious beings, resulting in a much greater variety of entities, experiences and extremes". If we were all perfectly adapted to our niches, then we would end up like the shark—so perfectly adapted to its niche in the oceans that it has not evolved for over 60 million years. And there is no question that us humans are evolving more rapidly than ever through our consciousness, culture, socioeconomic systems and technologies. In a nutshell, "The Trouble with Paradise" lies in our imperfect adaptation to our increasingly rapidly evolving, ever more complex niches. And therein also lies its genius.

You might well ask how such a potentially dry topic could be fun, useful and stimulating. There you are, reading these words, wondering exactly how I am going to get out of that one. Well, it turns out that an evolutionary philosophy can be made fun by including some personal tidbits that enliven the pages. And that "The Trouble with Paradise" proves to be one of the most useful things you might ever read, because it deals directly with the everyday situations you find yourself in, and provides a way of feeling good about yourself and enjoying life, the universe and everything, while also empowering you to deal more effectively with any situation you find yourself in.

So, what exactly are we going to cover in these pages?

In Chapter 1 "Why Some Like it Hot and Opposites Attract", we explore the ways in which our universe has, since the very beginning of time, been producing ever more complex, conscious entities and beings including ourselves through fusion, synergy and synthesis, and how an ever more <u>diverse range of beings, experiences and extremes</u> arise from that. We also learn how conflict, compromise and synergy are built-in features of such diversity, yet also transformations of each other, and how ecosystems evolve in nature through the balancing of these three primeval forces. The chapter finishes with reflections on the current clashes between ecology and economics and how they might be better understood in human terms.

In Chapter 2 "Great Questions, Big Thinkers, Many Answers", we explore how the minds of men (mainly-apologies to my feminist readers), have conceived of the nature of existence, and attempted to answer the eternal questions about life, the universe and everything, since the dawn of recorded history. From the earliest philosophers, through the founders of the major religions to the scientists and leading edge thinkers of the past few centuries, we follow the intertwining, ever evolving journey of faith and reason. We also revisit my very own humble beginnings going back to my exiled ancestors who chose the fairest Cape as their ticket to paradise three centuries ago.

In Chapter 3 "A Good Life, in the 21st Century?", we consider how the rapid acceleration of industrial and information technologies in the past few centuries has profoundly changed what it means to be fully human and to live a good life. Thrown into the mix are some reflections from my own life as a both a proponent of the benefits of such technologies, and also an activist battling some of the worst excesses of the "system", while struggling with my own personal troubles in paradise. The chapter ends with my own personal search for an "Oasis" where I can reinvent my life and create a small corner of paradise that might, in its own humble way, show what is possible and give others a "taste of paradise".

In Chapter 4 "The Trajectory of our Species", we revisit 14 or so billion years of Big History from the mighty mitochondria to multicellular creatures, and discover how our emergence as a species "Out of Africa" guaranteed the eternal success of the BBQ, among other things. From the origins of our earliest proto-human ancestors, to the flowering of our artistic impulse giving rise to the magic of the sound of music, culture and technology, we see the glorious shifts taking place in our species from our egocentric beginnings, through the rise of ethnocentric cultures, to the dawn of the worldcentric civilization that is emerging on our planet in the 21st century.

In Chapter 5 "Paradise began in a Garden", we discover why it is that we have a great deal to learn from gardens and gardening about the main principles of healthy living systems, and what this might mean if we managed our human systems like gardens and nature. As we learn to partner Creation in the re-creation of ourselves and the biosphere, we find that the successful future of our species might well lie in how soon we can create the alignment, resonance and coherence that will accelerate our conscious evolution as a species in a thriving planetary civilization.

In Chapter 6 "On the Origins of Heaven and Hell", we weigh in the balance the prospects for humankind to master the eight challenges posed by our complex world in which most of us are "in over our heads", and Munch's painting "The Scream" has become the symbol of our collective mental state. Shifting our responses to Hell on Earth is going to take more than a few doses of strong religion or the over-simplified atheistic response to religion, as we shall see. We then review some of the adaptive approaches deployed by modern humans to the challenges of the 21st century, and how perspective management might enable us to move beyond our exteriors to better deploy the four core intelligences that shape those perspectives. We finish off with a brief overview of the paradise creation business, and how we can use Big Psychology to build on what we've learned about Big History, to accelerate our own evolution in a paradise creating direction.

In Chapter 7, "The Arc of Human Development in the 21st Century", we dig a little deeper into the great shift that is unfolding in these turbulent yet opportunity-rich times, using the lens of developmental psychology and integral models of human development. After navigating the nine stages in the development of human consciousness, culture, socioeconomic systems and technology, we examine how the transpersonal stages of development from the integral stage and beyond set the scene for a great leap forward in human evolution in this century. The implications of this great leap forward for several fields is explored, from biology, medicine and longevity through consciousness and the mind, to the demystification of paranormal phenomena, the catalyzation of distributed spirituality and human flourishing, open education and innovation and the embedding of thriveability into green growth economies.

In Chapter 8, "New Answers to the Age Old Mysteries of Life, the Universe and Everything", we survey some of the latest thinking and approaches that are shifting our current scientific and social worldview, and contrast that with the twenty world changing discoveries from fire to the Higgs boson. We look again at the scientific method's role in all of this, and broaden the ways in which scientific investigations can be applied to non-material phenomena, in a variety of fields from physics and cosmology to the life and social sciences, consciousness research and metaphysical questions.

In Chapter 9, "Overview of a Transpersonal Learning Journey", I share how my own personal journey through life illustrates the main developmental and psychological models encountered in these pages. We end with an overview of the ten key propositions that summarize the practical consequences of the latest scientific breakthroughs.

Chapter One

Why Some Like it Hot and Opposites Attract

"The mind is its own place, and in itself, Can make a heav'n of hell, a hell of heav'n."

John Milton, Paradise Lost

Male/female.
Electron/proton.
Young/old.
Matter/antimatter.

We live in a universe where opposites attract. This profound truth is almost a cliché in relationships with what we interestingly refer to as the "opposite sex", and even truer when we zoom back to the moment our universe was born in a big bang 13.75 billion years ago. Matter and anti-matter started colliding and fusing, with matter being enough in the ascendant that we are able to talk of things such as gravity and toast dropping on the floor jam side down. This is all given to us by matter outweighing antimatter by one part in 30 million, making the atoms that make up the book you are holding in your hands, just possible, with the odds of a miracle.

379 000 years or so after the Big Bang, the opposite forces of positively charged atomic nuclei and negatively charged electrons combined to form the most common element in the universe: hydrogen. A small while later, give or take a few million years, large clumps of hydrogen began to form gas clouds and galaxies, and stars were born, and that is where the real magic begins.

Temperatures are so infernally hot inside a star that hydrogen atoms begin to fuse into bigger and bigger atoms, in a process appropriately named "fusion". And it is in this process that the very building blocks of life are created. Almost 99% of the mass of the human body is made up of six elements forged in the heart of stars: oxygen, carbon, hydrogen, nitrogen, calcium, and phosphorus.[i] You are, literally, made of star stuff.

Even though stars do get very hot, they are not hot enough on their own to produce the heavy metals we commonly know and love as jewelry, aircraft and serious weaponry. Things you can put on your fingers and wear around your neck and other seductive places emerged in the heart of even hotter events: the explosion of giant stars known as supernovae. The heaviest element regular stars can produce is iron. For more durable bangles and baubles things need to get *really* hot, not just the measly six million degrees centigrade that we would experience at the center of our sun.

In the few seconds after a supernova explodes, temperatures can reach anywhere between 350 billion to a trillion degrees centigrade. Think of this next time you sip a scalding hot cup of tea which reached a maximum temperature of 100 degrees centigrade in your kettle shortly before you prepared your brew. It takes a very great deal of energy indeed, to produce precious metals.

Supernovae are extremely luminous and cause a burst of radiation that often briefly outshines an entire galaxy. Over the few weeks of its existence, a supernova can radiate as much energy as the Sun is expected to emit over its entire life span. The explosion expels much or all of a star's material at 10% of the speed of light, driving a shock wave into the surrounding interstellar medium. This shock wave sweeps up an expanding shell of gas and dust called a supernova remnant.

And that is where cool jewellery comes from, as well as nuclear materials such as uranium. We get to see a supernova every fifty years in our Milky Way galaxy, though they occur frequently throughout our universe. A beautiful spectacle indeed, though one does need to be a few million light years away not to be blown away by the shock waves and radiation.

But I digress.

Incredible new properties emerge when the synthesis of opposites occurs. Hydrogen and oxygen, for example are both gases which occur naturally in our atmosphere, and they are nebulous. Indeed, we can see straight through them and breathe them in and out of our lungs. Put two molecules of hydrogen together with one molecule of oxygen however, and you get H2O, or water, as it is more commonly known. As we all now appreciate, the most obvious thing about water is that it is WET.

Wetness did not exist before two H's and an O combined into water. And we definitely cannot breathe H2O in and out of our lungs. Wetness is what scientists and philosophers call an "emergent property".

And so it goes on, up the great chain of being, from the earliest forms of life to troupes of elephants and human hives: new properties and capabilities continuously emerging. That is simply what our universe does, au naturel. Which is, to some extent, what this book is about: how this great chain of being evolved from the simplest of substances into the highly complex global species that is now planning missions to Mars.

While that may be interesting to some, there is an even more compelling, one might say, urgent issue that needs to be addressed: why, if we humans, poised as we are at the pinnacle of the great chain of being, and who have never had it so good materially speaking, are also so stressed out and generally dissatisfied with being the summa cum laude of all predators?

My simple explanation of this rather complex phenomenon is this: as we evolve up the ladder of being, we find three things: the first, that the tension between the range of opposites in our lives and societies widens dramatically and often painfully; the second, that the better informed and more intelligent we are, the more humble we have to become about our ability to live meaningful lives and to change anything, even ourselves; and consequently, thirdly, that the cost of gaining the simplicity the other side of complexity can rise very steeply if we do not align ourselves and our lives well.

Let us begin with opposites.

While opposites attract, and produce interesting new properties and capabilities when they combine, they also create tensions at each new level of being. Nature resolves most of these tensions through the laws of physics, chemistry and biology, and the outcomes are, as we increasingly and happily discover, rather predictable. But when we get to psychology, economics, culture and social affairs, we find that things are rather a mess—a great, big, mess. Such a mess in fact, that, despite gazillions of people with the best of intentions, we are currently unable to stop ourselves warming our planet by 4 degrees (or more, in worst case scenarios), and driving millions of species into extinction, with seriously worrying consequences for us all.

Now, I want to assure you dear reader that my purpose is to entertain and educate, not to shock or distress, as I personally believe we will all be better off if we develop creative, enjoyable responses to the challenges we face. So let us continue in this positive vein.

There are three ways in which tensions between opposites can be resolved so that the energy underlying them is released. Conflict is the most obvious way this happens, because its sources and outcomes are generally more dramatic and sudden than the other two.

When tensions between two tectonic plates in the earth's crust reach a certain level, the plates slide against each other and release a huge amount of energy, called an earthquake. Such earthquakes in human affairs are both shocking and memorable—who can forget that gripping saga called "World War 2"? Even seventy years after the event, it is one of the first things schoolchildren learn about. Most history books are centered on wars and great battles.

As we discover on a daily basis, conflict is the very stuff of life, at all levels. We find it in sibling rivalry, between spouses and families, in competitive battles between firms, between politicians, between different national and ethnic groups and of course, most dramatically, between countries and cultures. Conflict costs lives, dollars, euros and yuan, and can create enmities that endure for centuries—we are still experiencing the consequences of the Crusades nine centuries later, for example.

Conflict is so damaging that we have developed entire systems of law and governance to regulate it, at all levels from disputes between neighbors to disputes between superpowers. At the economic level, we regulate competition, and try to create "level playing fields" which are fair to all competitors. We also release many of our primitive competitive urges through sports and competitions of all kinds.

In nature, conflict occurs most frequently at the adolescent stage of the development of an ecosystem or species. This is where the contention for resources it at its height, and also where the drive for diversity is at its peak. On the plus side, this ensures that the maximum number of species evolve to "search" the space for niches in their environment, which in turn ensures the maximum use of resources and acceleration of evolution. The downside is that when the system as a whole hits its limits to growth, there are mass extinctions. And that happens often in the unconscious evolution that takes place in nature, as we shall see.

Along with war, the other thing schoolchildren learn about at a tender age is the idea of evolution (unless they live in one of those gloriously backward places where God or Allah created the world six or so thousand years ago). When the theory of evolution was first being developed by Charles Darwin and others, England was in the grip of a succession of famines, which the economist Thomas Malthus warned was a sign that there were too many people for the amount of food available. In this fearful climate, Darwin's theory of evolution thus became strongly associated in the popular consciousness with another idea: "selection of the fittest".

Taken to its extreme this idea was used to justify all sorts of bad things being perpetrated by one group of people on another, from colonialism to the "robber barons" of late nineteenth century capitalism (and for that matter, early twenty first century capitalism), to the concentration camps of the Third Reich. Luckily for us all, evolution is a great deal more complex, beneficent and glorious than that, as we shall see.

So, if conflict is the most talked about form of the resolution of the tension between opposites, then the second most talked about way of channeling the energies underlying the tension, is compromise. Sadly, compromise is a weak and unstable form of resolution, as every badly written peace treaty or unhappy and temporary truce between sparring siblings or spouses, will attest. Just why is that, exactly?

Compromise usually develops as a result of a desire on the part of two or more parties to avoid conflict, or to "get along" and damp down conflict when the respective parties run out of energy for a fight. Like political candidates smilingly shaking hands with each other after a rancorous debate, outcomes in the compromise zone can only ever aspire to be a temporary truce between opposites, enabling them to sit uneasily alongside each other until the next shock or opportunity arises that sets them off again like firecrackers at a Chinese New Year's celebration. Compromises take much energy to enforce, and when the lid is taken off the pressure cooker, the steam will inevitably blow off.

So, if we don't like or want destructive conflict or temporary, unstable compromise, what other alternative is there? Ladies and gentlemen, may I introduce you all to the (drumroll please maestro): Synergy Zone.

Nature is built on and through synergies. In fact, anything that is enduring or sustainable is almost by definition, synergistic. We would not be here if it were not for synergies between opposites. There are many wonderful books about synergy, and it is not my purpose here to give you an exhaustive account of this magical phenomenon. Synergies can be very simple or incredibly complex. They come in all sizes. They are everywhere one can detect a pulse. In short, they are the least understood and most important aspect of our existence.

So let us at least develop a basic understanding of synergy before we discover how we can move gracefully from this unsustainable world we have created, into the bountiful thriving world that is possible the other side of synergy.

Let us begin with simple synergies. As every English child knows, Jack Sprat could eat no fat, his wife could eat no lean, so betwixt the two of them, they licked the platter clean. This is a nice example of a complementary synergy, where two opposites are more effective with resources than they would be on their own, where each of them would have left half a meal on the table.

Ecosystems in nature and business both work better when there are complementary synergies between species and firms. For example, one firm's waste becomes the food for another. Landfills of rubbish which produce the toxic greenhouse gas methane, can sell their waste (methane) to nearby businesses which need cheap energy. Bingo, no wasted methane and less global warming! (Even though the landfill is itself still a giant rubbish dump).

In nature, every species finds a specific niche which enables it to exploit very specific foods and environments without generating too much competition with other species. Even carnivores such as lions will generally only eat the weakest or sickest animals in a herd, and only when they are very hungry. Nature does not waste a single molecule.

Every child knows that birds and bees are attracted to pollinate flowers by the nectar each flower offers the pollinators. They also know that big fish eat little fish, and that big fleas have little fleas to bite them, and little fleas have lesser fleas and so ad infinitum. Leaves, animal feces and other dead organic matter become compost and nourish the soil. And so it goes for the hundreds of millions of species on our planet—not a molecule wasted.

Yet in our human world, we still have a long way to go to embed this level of graceful efficiency into our social and economic systems. Despite forty years of environmentalism, as the seventh continent of floating plastic and other garbage known as the mid Pacific gyre attests, we are still pumping unprecedented amounts of pollution into our ecosystems. As architect Bill McDonough, the co-inventor of the

"Cradle to Cradle" approach to sustainability and design points out, we need to begin thinking about redesigning our industrial and social systems so that one firm's waste is another firm's food. This trend is growing rapidly as companies seek to save money, energy and resources in their attempts to become more sustainable.

Beyond simple synergies, we can find more complex examples in our human world by looking at a village, town or city. Or a region of the world, like Catalonia, which is where Spain meets France and the grand snow-capped peaks of the Pyrenees meet the salty sweet waters of the Mediterranean Sea. From Barcelona to Perpignan along the coast, and further inland toward Zaragoza and Toulouse, we find this eco-region thriving today thanks to Catalonia having largely escaped the worst effects of the industrial revolution and modernism. So what are they doing right that we can learn from and celebrate?

In the beginning, the Catalans say, God created Catalonia, and he saw that it was good. In fact, it was too good. Not only did it boast the snow-capped peaks of the Pyrenees, but those majestic peaks descended in a riot of natural beauty into the azure waters of the Mediterranean Sea. To top it all, in the fertile soil on the plains between the mountains and the sea, everything from Seville oranges to olive trees and exotic palms sprouted with wild abandon under the golden Mediterranean sun.

So, the story goes, in order to make things a little more balanced, God created the Catalans.

After the French won possession of northern Catalonia from the Spanish under the Treaty of the Pyrenees in 1659, this apocryphal story was later updated by the French philosopher Jean-Paul Sartre, who summed it up even more succinctly when he confirmed that: "Hell is other people".

For several decades I have been searching for new evidence to disconfirm this apparently universal truth, but to no avail. Despite the fact that we apparently live on one of the most beautiful planets in the universe, we appear to be surrounded on all sides by difficult neighbors, greedy governments, fanatics of all kinds, snakelike colleagues, irritating or embarrassing family members and the many complicated paraphernalia of modern life including packaging designed by clever machines which only another machine or a sharp pair of scissors can open.

In fact, there are several laws to describe this effect, named after an equally apocryphal series of fellows named "Murphy" and "Sod", giving us the modern expressions: "Murphy's Law" and "Sods Law", which both state the following:

"If something can go wrong, it will, and at the worst possible time".

There are hundreds of variations on these important laws, which for some reason have not yet been included in any school curriculum, though they are certainly more powerful than the laws of gravity, the weak and strong forces and electromagnetism combined. For example:

"The chance of the bread falling with the jam side down is directly proportional to the cost of the carpet."

Searching for synergy is all very well, but when your bread keeps on falling jam side down on the Persian rug, one is either very clumsy or there is something one has not properly understood yet, or perhaps both.

Outcasts from Eden: The Dismal Sciences

"And unto Adam He said: 'Because thou hast hearkened unto the voice of thy wife, and hast eaten of the tree, of which I commanded thee, saying: Thou shalt not eat of it; cursed is the ground for thy sake; in toil shalt thou eat of it all the days of thy life. Thorns also and thistles shall it bring forth to thee; and thou shalt eat the herb of the field. In the sweat of thy face shalt thou eat bread, till thou return unto the ground; for out of it wast thou taken; for dust thou art, and unto dust shalt thou return."

Genesis 3: 17-19

"I love my mother . . . . One thing she has taught me is that the things of most beauty in the world are worth suffering for. We must be what it takes to have those things that are beautiful, and to achieve any such things requires a lot of help. Any involvement with others in a world that keeps us usually so alone is part of the greatness of Everest or of any other of life's masterpieces of experience . . . . [ii]"

Pat Ament, Mountaineer

The root of the word for "eco" in both ecology and economics is the Greek word for "house" or "household". Women have, throughout history, been the pre-eminent housekeepers and managers of the household economy. If you are lucky enough to have had a good mother, then you will remember the sense of healthy order, being loved, pleased, entertained, protected and cared for by a well-run household led by a natural economist. Yet the nurturing, feminine side of economics has been overrun in the past few centuries by a macho discipline led by cold and calculating people who believe that we are all "rational" creatures making choices purely on the basis of maximizing our own selfish-interests. Ecology and economics have spent a century or more increasingly at war with one another, as our general wellbeing as a species has diverged sharply from our narrow interests as producers, consumers, bosses and employees.

So we end up in a world run by the richest 1% for themselves, by themselves, despite the wishes, votes and needs of the 99%, simply because the system has been bought and rigged by big money, big business, big media and big government. A world programmed for tension, conflict, misery, acting like a runaway virus destroying its host. Economics as we know it apparently has a deeply sublimated death wish.

Brian Eno, the musician, artist and producer of albums by U2, Cold Play, Paul Simon and many others, is a great fan of ecology. Like him, I believe that ecology is probably one of the most important ideas to emerge in the past 150 years, though I would put it on a par with the equally powerful idea of evolution. Ecology, together with systems thinking, give us the opportunity to experience a sense of interconnectedness with nature and our fellow human beings, without having to first get all mystical.

Eno puts it this way:

"Beginning with Copernicus, our picture of a semi-divine humankind perfectly located at the center of The Universe began to falter: we discovered that we live on a small planet circling a medium sized star at the edge of an average galaxy. And then, following Darwin, we stopped being able to locate ourselves at the center of life. Darwin gave us a matrix upon which we could locate life in all its forms: and the shocking news was that we weren't at the center of that either—just another species in the innumerable panoply of species, inseparably woven into the whole fabric (and not an indispensable part of it either). We have been cut down to size, but at the same time we have discovered ourselves to be part of the most unimaginably vast and beautiful drama called Life."

There are many possible responses to this "loss" we feel at not being the center of the universe. Since the beginnings of the first Renaissance some 500 years ago, the 57 varieties of our all too human response have included:

Denial—fundamentalists who believe that their faith is the only truth, and that all others, (including the ideas of evolution, ecology and science), are fundamentally in denial about this shift of mankind from the center of the universe to being a tiny part of a massive, apparently uncaring system;

Modernism—modernists worship progress, and accumulate achievements and possessions to validate their sense of their own importance and significance in the universe. There is a great deal of hope in projects such as the space programmes of various nations, the development of new and better technologies which makes our lives more comfortable, longer and healthier, and the emergence of democratic systems of government around the world. Strongly associated with "Skeptics".

Existentialists—existentialists combine varying degrees of fatalism and nihilism, and their attitude can be summed up in the phrase: "Plus ca change, plus c'est la meme chose". So what if we have advanced technologies—they are used as often to kill large numbers of people as they do good, they say. Yes, we have made some progress materially as a species, but as Jean-Paul Sartre put it: "What will future historians say of modern man—that he read the newspapers and fornicated?" Strongly associated with "Cynics".

Mystics—mystics today come in many flavours, from old hippies escaping the rigours of modern life, to earnest new agers who believe in developing themselves and sign up for the latest new trends and fads as they reveal new secrets of life, the universe and everything. The "Secret" movie is a good example of the narcissistic edge of the mystical "you can have it all, and deserve it all" fringe. Oh, and don't forget to buy the book and the DVD so you can find out that the secret to having anything you want is to "quaff" it into being through the sheer power of your new age mind.

Activists—thankfully, there are also people who believe in going out and righting wrongs to make the world a better place for all of us. They are the activists. At their best, they are being the change they seek in the world, and making a real difference to others—social entrepreneurs, eco-village pioneers, protesters, political activists, environmentalists, fair-trade activists, and so on. At their worst, they can be violent anarchists who undermine our existing dysfunctional systems, without putting anything else in their place.

Terrorists—very angry activists and fundamentalists often seek to achieve their ends through desperate measures. They say things like: "One man's terrorist is another man's freedom fighter".

On its own, terrorism generally makes things worse, though it does relieve the extreme emotional tension felt by the terrorists for a few brief moments before they blow themselves and others up. Occasionally, a Nelson Mandela will blow up a few power pylons to show the government they mean business, but most of the time terrorists do not possess one percent of the nobility of the Gandhi's and Mandela's of this world. They are scared, often cold blooded killers manipulated by cynical interests behind the scenes . . . who also deserve our compassion even as we seek to eradicate terrorism and its causes.

Although our brains, nervous systems and bodies require hundreds of hormones and neurotransmitters to function well, there are a few key substances which generate some of the greatest polarities in human nature. Testosterone is the hormone responsible for turning a female fetus into a baby boy, and then a man. Testosterone acts as the agent for individuation and independence, fuelling our curiosity and exploration. Testosterone is also released during aggression, sharpening our senses and readiness for action. Oxytocin, on the other hand, is the hormone responsible for bonding us together—whether a mother to a child, close ties between the members of a family, or the male /female bond in romantic relationships.

In our modern world, the stereotype is that Mom gets to run the house while Dad goes out to work. Both Mom and Dad do this thing called "management", but Dad's job is "more important" because he gets to manage a business, a country, a war, or something, which is usually bigger, and for guys bigger is better. Yet management is a tertiary skill—a method, not a value, even though we apply it to every domain as if it were the ideal of our civilization.

The caring, sharing, cooperative gift economy (or love economy, as Hazel Henderson calls it), was always about the unpaid 50% of all productive work mostly by women in all countries, including raising children, caring for households, serving on school boards, growing your own food, building your own houses and a thousand other invisible daily tasks. The contributions of this Love Economy never appeared on balance sheets and, thus, were also missing from GDP national scorecards. This problem became central to today's reforms of the GDP. In villages in developing countries, economists and analysts now realize that the official GDP figures have ignored the Love Economy which could total as much as 90% of all productive work and livelihoods in these areas.

If there is a mother of all "blind spots" in our emerging global civilization, this is it. In my own experience of life, the most excruciating places I have ever been to have been where management combined with macho dominate—they are also the most dangerous. The military (anywhere), the bad bits in the south of Spain, Naples, African townships, ghettoes in New York, LA and Miami—scary places indeed. Yet one can wander around Marrakech after midnight and feel perfectly safe in the pre-modern Arab world.

Testosterone is a deadly chemical in the wrong hands, and it is our culture that determines just how dangerous that stuff can be. When testosterone is used for sports, exploration and discovery, it is a wonderful thing, but if it is cooped up with nowhere to go, it becomes toxic and often lethally explosive. And in business and management an excess of testosterone has been, to put it simply, deadly. As planet earth has been discovering for over 300 years, where we have magnified the impact of testosterone a thousand fold through machines that enable both construction and destruction on a global scale. It is no wonder the 20th century was the deadliest in history, with hundreds of millions of deaths due to wars in the name of control and conquest, run by technically proficient "managers".

The irony is that while managers (whether bureaucrats, business executives or officers in the military or police) seek to control, the techniques of management themselves operate to distance these managers from what they are controlling, thereby cutting them off from the very connections and information needed to be effective. The "bosses" and their minions always succeed in distancing themselves from those they are controlling, while attempting to apply pseudoscientific methods of coordination and communication to achieve their goals. In such a context, for example, a little "collateral damage" (American military code for: "innocent people killed or injured as a result of attempts to control through the use of force)", is deemed acceptable by those in charge.

The world had seen this from time immemorial, and in the hundreds of wars big and small, fought between 1900 and 2000. Even in times of relative peace during the Cold War, the world watched as authoritarian regimes in the USSR, South Africa, Spain, Eastern Europe, South America, China and South-East Asia inflicted collateral damage on the innocent while crushing their opposition with the force of arms, terror and torture. The USA also had blood on its hands from its many and varied crusades against first "communism", then "terrorism". The infamous management techniques introduced in the Pentagon by Robert McNamara during the Vietnam War were able to count body bags, bombs dropped and enemy casualties with great precision, but unable to win the war or the peace.

Luckily for civilization and nature, the hand that rocks the cradle eventually rules the world. In other words, mothers shape the values of each and every generation that has ever walked the earth. And it is our mothers who teach us to clean up after ourselves, to "waste not want not", and generally to be nice to others and to respect them and their stuff. When we harmonize and align testosterone and oxytocin and the mind sets that produce them, we find wellbeing and happiness become real possibilities rather than Utopian dreams. Well-adjusted individuals, families, enterprises, cities, countries and planets depend on this kind of balance and alignment to exist at all.

There should be no fighting in Paradise, just enough creative tension to make life enjoyable, interesting and sustainable. Homo sapiens has produced well-functioning civilizations thousands of times throughout its history, everywhere on the planet. Yet for every heaven on earth we have produced, there has also been a shadow side that became a living hell for those deprived of the light. And every human civilization that ever existed has eventually become extinct, as the energy and resources required to sustain each civilization were depleted through excessive growth. At the peak of each civilization, the elite became blind to the consequences of its impact on the larger systems that sustained it.

Can we escape the same fate through renewable technologies, sustainable ways of living and more advanced levels of consciousness and awareness? It all depends on how well we can get to grips with the deeper sources of the Trouble in Paradise.

Chapter Two
Great Questions, Big Thinkers, Many Answers

The fundamental cause of trouble in the world today is that the stupid are cocksure while the intelligent are full of doubt.

Philosopher and Mathematician Lord Bertrand Russell

As a species, we have been fascinated by the incredible mysteries of the universe around us since the dawn of human time. We can imagine a cave family staring up into the inky depths of a night sky unspoiled by modern lighting, sharing stories about how their world originated and what their role in this game we call life was. The Bushmen still hold their children up to the splendor of the Milky Way, and ask of the universe that their child become a star.

Creation myths abound in aboriginal societies. "Mother Nature" is a common anthropomorphized representation of nature that focuses on the life-giving and nurturing features of nature by embodying it in the form of the mother. Images of women representing mother earth, and Mother Nature, are timeless. In prehistoric times, goddesses were worshipped for their association with fertility, fecundity, and agricultural bounty. Priestesses held dominion over aspects of Incan, Assyrian, Babylonian, Slavonic, Roman, Greek, Proto-Indo-European, and Iroquoian religions in the millennia prior to the inception of patriarchal religions.

Algonquin legend says that "beneath the clouds lives the Earth-Mother from whom flows the Water of Life, who at her bosom feeds plants, animals and men". The word nature comes from the Latin word, 'Natura', meaning birth or character. In English its first recorded use, in the sense of the entirety of the phenomena of the world, was very late in history.

The pre-Socratic philosophers of Greece had invented nature when they abstracted the entirety of phenomenon of the world into a single name and spoken of as a single object: Natura. By the Middle Ages "natura", and the personification of Mother Nature, was very popular throughout Europe.

In the modern, western world, we still find many people clinging to our own version of this creation myth, where God created heaven, earth, nature and people in six days, saw that it was good, and then on the seventh day he rested, and handed the Creation over to us.[iii] The part of our western creation myth that is becoming a problem for us today, is the following:

"Then God said, "Let us make mankind in our image, in our likeness, so that they may rule over the fish in the sea and the birds in the sky, over the livestock and all the wild animals, and over all the creatures that move along the ground."

So God created mankind in his own image, in the image of God he created them; male and female he created them. God blessed them and said to them, "Be fruitful and increase in number; fill the earth and subdue it. Rule over the fish in the sea and the birds in the sky and over every living creature that moves on the ground."

Then God said, "I give you every seed-bearing plant on the face of the whole earth and every tree that has fruit with seed in it. They will be yours for food. And to all the beasts of the earth and all the birds in the sky and all the creatures that move along the ground—everything that has the breath of life in it—I give every green plant for food." And it was so.

God saw all that he had made, and it was very good. And there was evening, and there was morning—the sixth day."

Indeed, it was good, very good, for a very long time. So good in fact, that there are now more than seven billion of us Homo sapiens crowding into every corner of our over-extended planet. We were exceptionally fruitful, and multiplied exceedingly. Our dominion as Lords over nature has led to what now amounts to the sixth mass species extinction since life began on earth four billion years ago, and the prospect of an out of control climate system for the foreseeable future.

But let us not get ahead of ourselves. The Big Questions of Life abound, and still remain unanswered: "Why are we here?", "How did we get here?", "What is the good life?" and "Where do we go when we die?" Climate change, mass extinctions, peak oil, peak soil and other challenges may have sharpened our desire to answer these questions in new ways that will be more sustainable in the future, but even if we succeed in meeting these challenges, the Big Questions will always be with us.

What is the purpose of life, my life, your life, everyone's life? Oh well, I give up, many say, that is too hard to answer, so why bother asking? Others cannot leave these questions alone, and have found out some very interesting things along the way. Life can be full of mystery and magic. Why are we here? Where did the universe come from? Six year old children come equipped with these questions: "Where does life on earth come from?; "Is heaven outside space and time in the universe?"; "What happens when you die—does your body stay on earth and your soul go to heaven?"; "Can I come back to earth as Bambi?". Children are very spiritual beings, before cynicism and peer group pressure begin to emerge as powerful forces in shaping what they believe life is all about. As adults, we often forget to consider or be amazed and humbled by the enduring mysteries of life.

Many great minds have already spent entire careers and working lives faithfully documenting the key "Big Ideas" and "Big People" that have attempted to answer the "Big Questions". In this chapter I will stand on their broad shoulders and share their essential insights with you. Luckily, unlike the galactic computer in "Life, The Universe and Everything", we may find some answers more useful than "42"[iv].

It is said that Philosophy asks questions that may never be answered, while Religion provides us with answers that may never be questioned. Hold onto your hats, ladies, gentlemen and LGBT's, as we fast-rewind to the first musings of our species on the Big Questions we have been asking for millennia, without getting much closer to an acceptable (let alone mutually agreed) set of Big Answers.

It is often said that knowledge is power, but it is always transient. Philosophy has taught me that self-knowledge is the only abiding form of true power. But where do we find self-knowledge? Self-knowledge is found within—and it defines your purpose, passion, highest possibilities & enables you to find your unique place in the world.

In order to do justice to such a weighty topic, I should start with a gentle reminder: most of us are generally unaware that we view the world from a very specific perspective. This usually stems from an education system faithful to the tenets of industrial capitalism, which like the production line and other inventions of the Industrial Revolution and the Enlightenment, is the product of an era in which the "Big Split" happened.

At this point, most authors would tell you that the "Big Split" is all the fault of a Frenchman named Descartes. In fact, they have named this problem the "Cartesian duality", because Descartes was a very clever guy who actually wanted to know how he could know anything, at all. To the point where he wanted to understand how could even know he actually existed, which is not something we normally get much time to think about in the daily hustle and bustle of increasingly complex and challenging 21st century life.

Essentially, most philosophers and spiritual types want to know two things, which are in turn probably the "Biggest Questions of All": "Why are we/am I here?", and "Where did we/I come from before I was born, and where will we/I go when we die?".

Perhaps the third Big Question in order of importance would then be:

"Are my conscious awareness and my subconscious intuition purely the result of physical phenomena in my body and brain, or are they a product of an interaction between my body and brain and another level/s of existence/ dimensions?"

Or put more simply, am I a human body that produces consciousness and imagines it might have a spirit, or am I a spiritual being who just happens to inhabit a human body?

Cognitive scientists call this the "Hard Problem" of Consciousness, and it IS hard precisely because of the terms in which it is framed, which always seem to return to a form of being forced to decide whether we are simply a purely material phenomenon, which hard science will eventually explain, or whether, indeed, there is more to matter, energy and information than meets the eye?

Descartes tried to resolve this by suggesting that the soul and the body were of completely different substances, yet interacted in the pineal gland, thereby "resolving" the dispute between the logical scientists who insisted that religion was just a crock of superstitious baloney, and the theologians who insisted that God and the soul were as real as the atoms and energy we are made of.

Reductionists like Daniel Dennett maintain that the Hard Problem is really a bunch of simple problems stuck together, and as science resolves each simple problem, there will eventually be nothing left to explain about consciousness: problem solved (someday)!

Non-reductionists argue that consciousness is an inherent, irreducible part of nature. The psychophysical branch of non-reductionists believes that the hard problem can only be solved through a psychophysical theory that includes fundamental laws. Roger Penrose and Stuart Hameroff argue that consciousness is indeed a quantum phenomenon, while Ervin Laszlo and others argue that consciousness is simply a reflection of the fact that the universe is made of information, and we are information aware of itself. The latter view is not that far off the hard science version of "M-theory", where physicists posit a version of string theory that gives rise to multiple universes, and where these universes are essentially made of information.

Then there are the Mysterians, who argue that we are unable to solve the hard problem because of the nature of our minds being limited by their very physicality, derived from the fact that we invent theoretical concepts by extending the initial concepts we form from our perception of macroscopic objects. Mysterians use the analogy of human beings trying to understand consciousness being like a rat trying to understand calculus—we are simply "closed" off from solving the problem as our minds were not constructed to solve it in the first place.

Most educated people today, especially in the so-called "western world" (and increasingly in the "fast developing world" which now includes countries like Brazil, Russia, India, China, South Africa or the "BRICS"), have been taught that science and religion are total opposites: in science we get to "know" things through a carefully peer reviewed procedure using the scientific method. In religion, we are told we have to take things "on faith"—i.e. we have to believe certain things to belong to this or that religion, without questioning why. Thus, knowledge and belief are arrived at through two very different, contradictory processes.

But let's dig a little deeper into the scientific method. To "be scientific" and rational, one cannot trust anything except "facts". Said facts are arrived at through the scientific method, which requires us to step back from whatever we are observing to begin with, in order to be "objective" about it. We then form a hypothesis about the phenomenon in question, and then test whether this hypothesis (or "theory", as it is often called), is true, by conducting an experiment or two. Providing we have followed scientifically acceptable experimental methods, we then arrive at an answer of sorts: either our theory correctly explains the observed phenomenon, in which case it is true, or it does not, and our theory turns out to be false.

This is one of the reasons scientists find the Hard Problem a tough nut to tackle, as it presents very few ways (if any at all), in which it can be tested through an experiment, so it is not possible to say whether or not consciousness is a purely physical phenomenon in a purely physical, material universe (what we might call the "Madonna" approach, following her song "Material Girl"), or whether consciousness is indeed a kind of gateway to other worlds which we can only intuit by going "inside", and exploring the unexplainable experiences that have been documented over thousands of years by many thousands of sages and experimenters. And the answer to this problem is definitely not on Wikipedia, nor may it be anytime soon.

Science, though useful, appears to explain precious little when it comes to matters of ultimate concern—e.g. why am I here, what happens after I die? The fact that physicists have had to invent "Dark Matter" to explain their cosmological theories, is just the beginning of the problems physicists have with explaining anything and everything. Which may explain why many of the most respectable physicists have ended up being intelligent mystics. A little humility is definitely in order for those who claim to know a great deal more than people who have spent a life's work grappling with these questions .... although we may now safely be wryly amused at some of the earlier attempts our species has made at answering the ultimate Big Questions.

Despite thousands of years of thinking and a very long line of philosophers in the west (dating from Thales of Miletus in southern Asia Minor, born in 624 BCE; Pythagoras who showed up in 570 BCE and Socrates who was born in Athens in 470 BCE), we appear no closer to any useful, rational description of how to live "the good life".

Early philosophers asked questions like: "What is the world made of?" and "What holds the world up?" Thales thought the whole world was made up of a single element, water. Heraclitus said that everything is a flux—immortalized by his statement that one cannot step into the same river twice.

Pythagoras studied mathematics and philosophy, and believed the universe had an innate order which could be expressed in mathematical terms. The physicists called "string theorists" are still continuing

Pythagoras' efforts toward a correct fundamental description of nature almost three thousand years on.

An intriguing feature of string theory is that it involves the prediction of extra dimensions. According to Wikipedia:

"The number of dimensions is not fixed by any consistency criterion, but flat spacetime solutions do exist in the so-called "critical dimension". Cosmological solutions exist in a wider variety of dimensionalities, and these different dimensions are related by dynamical transitions. The dimensions are more precisely different values of the "effective central charge", a count of degrees of freedom which reduces to dimensionality in weakly curved regimes."

One such theory is the 11-dimensional M-theory, which requires spacetime to have eleven dimensions, as opposed to the usual three spatial dimensions and the fourth dimension of time. The original string theories from the 1980s describe special cases of M-theory where the eleventh dimension is a very small circle or a line, and if these formulations are considered as fundamental, then string theory requires ten dimensions.

But the theory also describes universes like ours, with four observable spacetime dimensions, as well as universes with up to 10 flat space dimensions, and also cases where the position in some of the dimensions is not described by a real number, but by a completely different type of mathematical quantity. So the notion of spacetime dimension is not fixed in string theory: it is best thought of as different in different circumstances.

So much for a truly useful answer to what the universe is made of, how it works and where it will end (if it does end). Will it take another thousand years to find an answer? No one knows for sure. I have always admired the physicist Richard Feynman, and after reading a series of rather complicated books in my attempt to understand quantum physics and the general and special theories of relativity three decades ago, the only sane response I could make was to write the following piece of doggerel:

Quark, Quark

(with an apology to pions and mesons)

It's quite a lark, being a quark
after dark
in the vacuum of the particle accelerator
it leaves many bubbles for its troubles.
What drives the muon, hadron and lepton?
And what would we do without subnuclear glue?
Things would fall apart
Without a fearful hand and eye,
To frame theories of super symmetry.
Protons, neutrons and baryons, the heavies in the pack
They've got the weight the tiny neutrinos lack
In the eightfold approach to the universe, there is no room for doubt
We live with quantum uncertainty and relativity is out!
Yet, we keep our feet on the earth
Gravitons and gravitinos make sure of that, they also stop
The world being flat.
What would McCavity have made of supergravity, no doubt
Some strange depravity?
Or the interquark chromodynamic force, which leads to a divorce,
Between acolytes of grand unified theory and non-believers,
of course.
Does the ultimate structure of the universe really matter?
Or is it just a lot of patter, by physicists,
madder than the proverbial hatter?

One cannot really say much more about the usefulness of the centuries of attempts by physicists to crack the code of our universe. Not only that, it is getting so much more expensive to find magical particles like the Higgs Boson, otherwise known as the" God" particle, because finding this little beast would help unite the worlds of the infinitely small (quantum mechanics), and the humungously large (the theory of general relativity).

Or so the research proposals that justified the ten billion euros and counting on the world's largest particle accelerator at CERN deep under the Swiss Alps, are most likely to have pointed out. Of this I am certain—after dozens of years, tens of billions of euros, and miles of public relations and media hype, the discovery of the "God particle" will change nothing of any importance. You can quote me on that[1].

What I am actually waiting for is the warp drive that anti-matter research has been promising for so many years, so that I can intergalactically go boldly where no man has gone before, like those brave heroes in Star Trek, for the price of a Ryanair ticket to London. Surely, you would agree, that is not asking too much of a return for us ordinary folks from the trillions spent by physicists worldwide out of our taxes in the past century?

What's It All About, Alfie?

Who will tell whether one happy moment of love or the joy of breathing or walking on a bright morning and smelling the fresh air, is not worth all the suffering and effort which life implies.

Erich Fromm

Never abandon a theory that explains something until you have a theory that explains more.

John McCarthy

During Classical times and the Middle Ages, the prevailing model of how our world worked was that of the "Great Chain of Being": a pyramid with God at the top, Man a close second and, sharply separated below,

---

[1] In early July 2012, scientists at CERN announced that they now believe they have found the traces of the Higgs Boson, with a certainty of 99.99999% (five sigma). And, true to form, nothing at all has changed. Of course, they need to do "more research", whatever that means, but all we can know for sure is that it will cost all of us taxpayers a bundle.

a teeming mass of life and matter beneath. In that model, information and intelligence flowed in one direction only—from the intelligent top to the inanimate bottom—and, as masters of the universe, we felt no misgivings exploiting the lower reaches of the pyramid.

The classical worldview started with the "Big Men" who generated the "Big Ideas" on which we premise much of our thinking today. In the 21st century, we may find this a little quaint, as we recognize that countless minds contribute to a river of innovation. Although we still celebrate and admire the most conspicuous of such innovators we see them now both as sources and outpourings of that mighty river of co-creation. As we will see later, this has major consequences for the way we think about social design, governance, enterprise, crime and punishment, education and learning, culture and science, and why each of us is here in the first place.

Socrates was focused on this more practical agenda. He believed we needed to know how to conduct our lives and ourselves. His urgent questions were:

What is good?
What is right?
What is just?

If we preserve our integrity, Socrates believed, no real, long-term harm could come to us. He also believed that no one knowingly does harm, and that if we could only know accurately the answer to questions like "What is justice", then we would be bound to behave more justly. Plato's dialogues took the work of Socrates forward along the ethical lines with which he began, then began to move toward natural philosophy and mathematics. Although Plato was a faithful follower of much of what Socrates had taught him, he disagreed that virtue is simply a question of knowledge. He very pragmatically stated that virtue requires that reason rule the irrational parts of one's soul, which also made him the first "Freudian" by making a distinction between the conscious and subconscious minds.

Plato is most famous for his belief that our world is a constantly changing reflection of another timeless and unchanging world of which our everyday world offers us only a glimpse. He also believed that our body is only a fleeting glimpse of a non-material us that is timeless and indestructible: our soul. For Plato, life's ultimate aim is to move beyond a superficial understanding of things and to know its ultimate reality, which is to be achieved through practicing detachment from the world and through philosophy.

Over two thousand years after Plato's death, the quantum physicist David Bohm arrived at a slightly more modern version of Plato's thinking via a lifetime spent researching the strange world of quantum physics. According to Bohm's theory, there is an implicate order (an eternal reality interior to us and therefore invisible) and an explicate order (the world of visible spacetime), both of which are contained in a holographic universe.

In the enfolded implicate order, space and time are no longer the dominant factors determining the relationships of dependence or independence of different elements. Rather, an entirely different sort of basic connection of elements is possible, from which our ordinary notions of space and time, along with those of separately existent material particles, are abstracted as forms derived from the deeper order. These ordinary notions in fact appear in what is called the "explicate" or "unfolded" order, which is a special and distinguished form contained within the general totality of all the implicate orders.

The implicate order represents a general metaphysical concept in terms of which Bohm claims that matter and consciousness might both be understood. Bohm proposes that both matter and consciousness enfold the structure of the whole within each region, and involve continuous processes of enfoldment and unfoldment.

For example, in the case of matter, entities such as atoms may represent continuous enfoldment and unfoldment which manifests as a relatively stable and autonomous entity that can be observed to follow a relatively well-defined path in space-time. In the case of consciousness, Bohm pointed toward evidence presented by Karl Pribram that memories may be enfolded within every region of the brain rather than being localized (for example in particular regions of the brain, cells, or atoms).

Bohm went on to say:

"As in our discussion of matter in general, it is now necessary to go into the question of how in consciousness the explicate order is what is manifest . . . the manifest content of consciousness is based essentially on memory, which is what allows such content to be held in a fairly constant form. Of course, to make possible such constancy it is also necessary that this content be organized, not only through relatively fixed association but also with the aid of the rules of logic, and of our basic categories of space, time causality, universality, etc .... there will be a strong background of recurrent stable, and separable features, against which the transitory and changing aspects of the unbroken flow of experience will be seen as fleeting impressions that tend to be arranged and ordered mainly in terms of the vast totality of the relatively static and fragmented content of memories.[v]

Bohm also claimed that "as with consciousness, each moment has a certain explicate order, and in addition it enfolds all the others, though in its own way. So the relationship of each moment in the whole to all the others is implied by its total content: the way in which it 'holds' all the others enfolded within it". Bohm characterizes consciousness as a process in which at each moment, content that was previously implicate is presently explicate, and content which was previously explicate has become implicate.

One may indeed say that our memory is a special case of the process described above, for all that is recorded is held enfolded within the brain cells and these are part of matter in general. The recurrence and stability of our own memory as a relatively independent sub-totality is thus brought about as part of the very same process that sustains the recurrence and stability in the manifest order of matter in general. It follows, then, that the explicate and manifest order of consciousness is not ultimately distinct from that of matter in general.[vi]

Spiritual pioneers and leaders have also struggled with similar questions since the time of Abraham in 1800 BCE. The life of Abraham takes up a good portion of the Genesis narrative from his first mention in Genesis 11:26 all the way to his death in Genesis 25:8. Although we know much about Abraham's life, we know little about his birth and early life. When we first meet Abraham, he is already 75 years old.

Genesis 11:28 records that Abraham's father, Terah, lived in Ur, a very influential city in southern Mesopotamia situated on the Euphrates River, about halfway between the head of the Persian Gulf and the modern day city of Baghdad. We also learn that Terah took his family and set off for the land of Canaan, but instead settled in the city of Haran in northern Mesopotamia, on the trade route from ancient Babylonia and the Mediterranean Sea about halfway between Nineveh and Damascus.

While Abraham was leading his people to the promised land of Israel and creating the roots of the Jewish nation, most of the rest of the world was worshipping multiple nature gods, just as many aboriginal tribes still do today. Abraham certainly did an excellent job establishing Judaism, though most of the credit for useful advice goes to Moses, who reputedly went up to the mountaintop and carved out the Ten Commandments in stone for his unruly followers as he led them out of slavery in Egypt, back to the promised land.

There is much useful practical advice here such as "Thou shalt not murder . . . steal . . . commit adultery . . . bear false witness against your neighbor . . . covet your neighbor's wife, ox or donkey . . ." and "Thou shalt honour your father and mother . . . observe the Sabbath day (which includes not working and not letting your slaves, children, livestock or resident aliens work). In overall terms, however, the Ten Commandments could hardly be said to be an inspirational document for modern generations, even if we would all find the odd slave, ox or donkey handy in taking the load off on a particularly bad day.

A little further to the south in Egypt a few centuries later, Akhenaten, a Pharaoh of the eighteenth dynasty of Egypt, was also developing his own religion which involved the exclusive worship of one God and the advocacy of universal values. His views represent an early expression of what was later championed by Judaism, Christianity and Islam.

Akhenaten's chief wife was Nefertiti, who has been made famous as the most "beautiful women in the world" by her bust in the Ägyptisches Museum in Berlin. Akhenaton was vilified by his successors for his neglect of the traditional religious cult and as a heretic in introducing monotheistic reforms. He was all but struck from the historical record. He remains a figure of great interest, however, and at least one writer describes him as the most original thinker of all the Pharaohs. The possibility that he made some contribution to the development of the three Abrahamic or Semitic faiths and their ideas, cannot be ruled out.

East is East, and West is West, shall the twain never meet . . . . ?

Oh, East is East, and West is West, and never the twain shall meet
Till Earth and Sky stand presently at God's great Judgment Seat;
But there is neither East nor West, Border, nor Breed, nor Birth,
When two strong men stand face to face,
tho' they come from the ends of the earth!

Rudyard Kipling—The Ballad of East and West

Meanwhile, on the other side of the world, Chinese philosophers began to stir. It all began with the Yi Jing (the Book of Changes), which was an ancient compendium of divination. The Book of Changes uses a system of 64 hexagrams to guide action, and is attributed to King Wen of Zhou (1099-1050 BCE), and reflects the characteristic concepts and approaches of Chinese philosophy.

The Tao Te Ch'ing of Lao Tzu and the Analects of Confucius (sometimes called Master Kong), both appeared around the 6th century BCE, slightly ahead of early Buddhist philosophy in Northern India and slightly after pre-Socratic philosophy in Ancient Greece.

Confucianism represents the collected teachings of the Chinese sage Confucius, who lived from 551 to 479 BCE. His philosophy concerns the fields of ethics and politics, emphasizing personal and governmental morality, correctness of social relationships, justice, traditionalism, and sincerity. The Analects stress the importance of ritual, but also the importance of 'ren', which loosely translates as "human-heartedness". Confucianism, along with Legalism, is responsible for creating the world's first meritocracy, which holds that one's status should be determined by ability instead of ancestry, wealth, or friendship. Meritocracy continues to be a major influence in Chinese culture today.

Throughout history, Chinese philosophy has been molded to fit the prevailing schools of thought and circumstances in China. Except during the Qin Dynasty, Chinese schools of philosophy were generally both critical and yet at the same time relatively tolerant of one another. Even when one particular school of thought was officially adopted by the ruling bureaucracy, as in the Han Dynasty, there was no move to ban or censor other schools of thought. Despite and because of the debates and competition, they generally cooperated and shared ideas, which they would usually incorporate with their own. For example, Neo-Confucianism was a revived version of old Confucian principles that appeared around the Song Dynasty, incorporating Buddhist, Taoist, and Legalist features within it.

Chinese philosophers and spiritual teachers were very flexible in their approach and focused on pragmatic issues such as how to rule the state, how to conduct one self and how to do the right thing. The ultimate goal of all of this was not to disturb the cosmic order or deviate from the "Dao", the set of principles that govern both nature and human beings. In this sense, Chinese philosophers were thinking along very similar lines to Aristotle back in Athens. Aristotle believed that the goal of life was happiness or "eudemonia", and that the science of ethics was both an investigation into how happiness can be achieved by human beings and how governments could enable citizens to live a full and happy life

Aristotle differed from Plato in that he believed all things are composed of matter, and that Plato's "form" is incapable of a separate existence to matter. As one of the first "reductionist materialists", Aristotle mapped out many of the basic fields of science as we know it, including psychology, physics, logic, economics, meteorology, rhetoric, biology and ethics. Much like many modern scientists, Aristotle really wanted to get to the bottom of things by breaking them up into their constituent parts and analyzing them. So, a lot of very useful, hands-on practical stuff emerged right there, along with the formation of the kind of thinking that kicked off the European Renaissance two millennia later: all thanks to one ancient, great Greek thinker.

After Aristotle's death, Hellenistic philosophy split into many different schools, each famous for a particular slant on the world. There were, inter alia, the Cynics, the Epicureans, the Skeptics and the Stoics.

The Cynics recommended enjoying life by reducing our needs to the minimum so that they could be effortlessly met. The most famous cynic was Diogenes, who lived in a barrel, and aggressively flouted convention. He is perhaps most famous for his short but rather pithy conversation with Alexander the Great which appears from all accounts to have gone something like this:

There lived a wise man in ancient Greece whose name was Diogenes. Men came from all parts of the land to see him and talk to him. When Alexander the Great came to town after having conquered half the entire world known to the ancient Greeks, he went to see the wise man. He found Diogenes outside the town lying on the ground by his barrel. He was enjoying the sun.

When Diogenes saw the great king he sat up and looked at Alexander. Alexander greeted him and said:

"Diogenes, I have heard a great deal about you. Is there anything I can do for you?"

"Yes," said Diogenes, "you can step aside a little so as not to keep the sunshine from me."

Alexander was no doubt a little surprised. Yet this answer did not make him angry. He turned to his officers with the following words:

"Say what you like, but if I were not Alexander, I should like to be Diogenes."

To put this into the proper context, by the age of thirty Alexander was the creator of one of the largest empires in ancient history, stretching from the Ionian Sea to the Himalayas. He was undefeated in battle and is considered one of the most successful commanders of all time. Alexander was tutored by Aristotle, so whatever else Diogenes had to say, which will go forever unrecorded, must have contained at the minimum a smidgeon of wisdom.

For the Epicureans, only atoms and space were eternal, and everything else changes. They believed that we can make the best of this life through moderation of our pursuits (much like the Buddhist "middle way"), and by withdrawing from public life and not getting involved in human affairs. Epicureans recommended that we seek pleasures that last, especially the basics such as food and drink, bed and friends (although not necessarily in that exact order but certainly in one or more interesting combinations). God was a distant figure who did not get involved in human affairs, hence could be ignored, they believed. Epicureans shared with the Stoics the simple belief that all tragedies can be endured by remembering that they will either kill you quickly or soon pass. Ah well, that's enough to cheer one up when one next gets a tad depressed.

The Stoics took this advice a little further, through their belief that reason was the highest authority. Therefore, the world we perceive is all there is, nature is governed by rationally intelligible principles (logos), and God is the spirit of rationality that imbues both ourselves and nature, as reflected in our minds and our self-awareness of the world and ourselves. (Again, a curiously parallel argument to the way the Buddhists describe us all as being a Buddha, or God within us—the eternal "Buddha nature" is in all of us and in everything else that exists).

Long before the classic existentialists such as Martin Heidegger, and the modern existentialists such as Jean-Paul Sartre and Albert Camus, the Stoics maintained that we get just this one shot at life, and that when we die, it is literally all over. No afterlife, no spirit, nothing—just decomposing matter and perhaps a tendency toward red hair and gout, or whatever distinguished you, showing up in your descendants.

Taking this elevation of reason to its logical conclusion, the Stoics recommended we accept all things without complaint, for indifference is the hallmark of reason. The logical corollary of this is that our emotions are unreliable and almost always lead to false judgments. In this spirit of ignoring your feelings, it was perfectly reasonable for the Stoics to take their own lives if things got so bad that one could no longer bear life's troubles with dignity and calm.

The Skeptics made the Stoics look like a bunch of over-emotional cry-babies. They refused to believe in anything, including their belief that they should not believe in anything. They argued, in what might have seemed quite reasonable at the time, that because there are always equally good arguments for both sides in any conversation, we should stop worrying about who is right or wrong and simply "go with the flow". Because every argument assumes its own starting point, as Timon pointed out, no ultimate ground of certainty can ever be reached in any logical proposition. Well, that settles that. Pass the hemlock!

Between the 6th and 4th centuries BCE, at much the same time that the Greeks were doing all of this heavy intellectual lifting and the Chinese were beginning their multi-millennium civilization, Siddhartha Gautama, who was born a prince and then forsook his inheritance for a monastic life, lived and taught in northeastern India. Commonly known as "the Buddha", (meaning "the awakened one" in Sanskrit), Siddhartha is recognized by Buddhists as an awakened or enlightened teacher who shared his insights to help sentient beings end suffering, achieve nirvana, and escape what they see as a cycle of suffering and rebirth.

The foundations of Buddhist tradition and practice are the Three Jewels: the Buddha, the Dharma (the teachings), and the Sangha (the community). Buddhism in modern times has helped many millions of stressed out urbanites in the development of mindfulness and the practice of meditation. Others go deeper into other areas of Buddhism involving the cultivation of higher wisdom and discernment, study of scriptures, devotional practices, and various kinds of ceremonies.

The world's most famous Buddhist at the time I write this is without any doubt the Dalai Lama, who exemplifies the moderate, tolerant nature of the "middle way" which is central to Buddhism. As he puts it: "My religion is kindness". Buddhists are not dogmatic about beliefs, but focused on how the mind creates stress and suffering, and how we can liberate ourselves from this suffering. Here then, is a very pragmatic philosophy and some very useful advice to us stressed out modern folks, as expressed in this excerpt of the words of "Atisha's Advice", from one of Buddhism's best-loved quotes given by the Indian Teacher, Venerable Atisha, in 11th Century Tibet:

- Since future lives last for a very long time, gather up riches to provide for the future.
- You will have to depart leaving everything behind, so do not be attached to anything.
- Generate compassion for lowly beings, and especially avoid despising or humiliating them.
- Have no hatred for enemies, and no attachment for friends.
- Do not be jealous of others' good qualities, but out of admiration adopt them yourself.
- Do not look for faults in others, but look for faults in yourself, and purge them like bad blood.
- Do not contemplate your own good qualities, but contemplate the good qualities of others, and respect everyone as a servant would.
- See all living beings as your father or mother, and love them as if you were their child.
- Always keep a smiling face and a loving mind, and speak truthfully without malice.
- If you talk too much with little meaning you will make mistakes, therefore speak in moderation, only when necessary.

Today, if you travel through Bhutan, Cambodia, Sri Lanka, Thailand and a small Russian republic known as "Kalmykia", you will find that both Tibetan and Theravada Buddhism are recognized as the official religion, though being Buddhists, they will not punish you for believing in someone else's religion.

Before we return to the account of my journey through the eye of the storm that was the first decade of the 21st century, and the question we opened this chapter with, there are several important belief systems we need to briefly explore to round out our short review of the major philosophies and religions that have pre-occupied our species since it could record its thoughts for posterity.

Believing We are More than Mere Chemistry Conspiring to be Human

"BIGOT—n. One who is obstinately and zealously attached to an opinion that you do not entertain."

Ambrose Bierce

Q: What's the difference between a fanatic and a zealot.

A: A zealot can't change his mind. A fanatic can't change his mind and won't change the subject.

Sir Winston Churchill

Are you religious, spiritual, atheist or agnostic? Or perhaps you've never thought about, or cared about it? And what is **it**, exactly, and why does it matter? Can that help you find out why you are here, and what really matters to you? And how might that help you, in your life?

Over 4.5 billion people identify themselves as having a religious affiliation, roughly two in three human beings. The world's largest religions that show increases that outrun birth-rate in their membership include Christianity, Islam, and Hinduism. Recent research[vii] shows that in Europe, roughly thirty percent of people say that religion is important or very important in their lives; over sixty percent of people in the USA say that religion is important or very important in their lives, while in virtually every country in the Middle East and Muslim countries in Asia, over ninety percent say that religion is important or very important in their lives.

Islam is currently the world's fastest growing religion, while in China, Vietnam and parts of Africa, Christianity is the fastest growing religion. With over 1.3 billion Muslims worldwide and over 2 billion Christians, the 1.3 billion other religious adherents are mainly accounted for by 900 million Hindus (80% of whom live in India), and 500 million Buddhists. The fastest growing religions in Europe and North America include Buddhism and variants of atheism and agnosticism.

Monsignor Vittorio Formenti, who compiles the Vatican's yearbook, quaintly said in an interview with the Vatican newspaper L'Osservatore Romano that "For the first time in history, we are no longer at the top: Muslims have overtaken us". The Monsignor said that Catholics accounted for 17.4 percent of the world population—a stable percentage—while Muslims were at 19.2 percent. "It is true that while Muslim families, as is well known, continue to make a lot of children, Christian ones on the contrary tend to have fewer and fewer."

Turning to our resident sage, Wikipedia, we find that religion is defined as:

"... a collection of cultural systems, belief systems, and worldviews that establishes symbols that relate humanity to spirituality and moral values. Many religions have narratives, symbols, traditions and sacred histories that are intended to give meaning to life or to explain the origin of life or the universe. They tend to derive morality, ethics, religious laws or a preferred lifestyle from their ideas about the cosmos and human nature.

The word religion is sometimes used interchangeably with faith or belief system, but religion differs from private belief in that it has a public aspect. Most religions have structures and rituals, including clerical hierarchies, a definition of what constitutes adherence or membership, congregations of laity, regular meetings or services for the purposes of veneration of a deity or for prayer, holy places (either natural or architectural), and/or scriptures. The practice of a religion may also include sermons, commemoration of the activities of a god or gods, sacrifices, festivals, feasts, trance, initiations, funerary services, matrimonial services, meditation, music, art, dance, public service, or other aspects of human culture."

Atheism, on the other hand, is, broadly speaking, the rejection of belief in the existence of deities. In a narrower sense, atheism is specifically the position that there are no deities. Most inclusively, atheism is simply the absence of belief that any deities exist. Given some of the drawbacks of religion over the centuries, It may come as no surprise that the number of atheists in the developed world is growing rapidly. Here is a short anecdote to illustrate the point:

"A journalist assigned to the Jerusalem bureau takes an apartment overlooking the Wailing Wall. Every day when she looks out, she sees an old Jewish man praying vigorously. So the journalist goes down and introduces herself to the old man.

She asks: "You come every day to the wall. How long have you done that and what are you praying for?"

The old man replies, "I have come here to pray every day for 25 years. In the morning I pray for world peace and then for the brotherhood of man. I go home have a cup of tea and I come back and pray for the eradication of illness and disease from the earth."

The journalist is amazed. "How does it make you feel to come here every day for 25 years and pray for these things?" she asks.

The old man looks at her sadly. "Like I'm talking to a wall."

In the twentieth century atheism became a major belief system in its own right with the ascendance of Marxist-Leninist systems of belief in Russia and China. In 1917, Vladimir Ilyich Lenin and his comrades kick started the Russian Revolution, and the Marxist-Leninist systems of belief eventually became the official state religion of the Union of Soviet Socialist Republics.

While decreed to be "the opiate of the masses", religion was tolerated. Although Soviet Marxism was incompatible with belief in the Supernatural, Communism required a conscious rejection of religion or else it could not be established. This was not a secondary priority of the system, nor was it hostility developed towards religion as a competing or rival system of thought, but it was a core and fundamental teaching of the philosophical doctrine of the Communist Party of the Soviet Union

Some 22 years later in China, Communism became the official ideology of the country. Quotations from the Works of Chairman Mao Tse-Tung, known as the Little Red Book, have since sold over 900 million copies, simply because during the latter years of the Cultural Revolution, it was compulsory to have a copy on you at all times for every Chinese person. The official ideology of the People's Republic of China was opposed to religion, and people were told to become atheists from the early days of the PRC's existence. During the Destruction of Four Olds campaign, religious affairs of all types were discouraged by Red Guards, and practitioners persecuted. Temples, churches, mosques, monasteries, and cemeteries were closed down and sometimes converted to other uses, looted, and destroyed.

Marxist propaganda depicted Buddhism as superstition, and religion was looked upon as a means of hostile foreign infiltration, as well as an instrument of the 'ruling class'. Chinese Marxists declared 'the death of God', and considered religion a defilement of the Chinese communist vision. Clergy were arrested and sent to camps; many Tibetan Buddhists were forced to participate in the destruction of their monasteries at gunpoint

Some atheists are irreligious, while others are spiritual: certain religious and spiritual belief systems, such as Jainism, Buddhism, Hinduism, and the Neopagan movements do not advocate belief in God or gods. Rates of self-reported atheism are among the highest in Western nations, although to varying degrees—for example: United States (4%), Italy (7%), Spain (11%), Great Britain (17%), Germany (20%), and France (32%)

Agnostics, meanwhile, are not sure whether God or gods exist. Agnosticism is the view that the truth value of certain claims, especially claims about the existence or non-existence of any deity, but also other religious and metaphysical claims, is unknown or unknowable. Thomas Huxley, an English biologist, coined the word in 1869. However, as we have already seen, several earlier thinkers promoted agnostic points of view, including Protagoras, a 5th-century BCE Greek philosopher, and the Nasadiya Sukta creation myth in the Rig Veda, an ancient Sanskrit text.

So, let us begin with the Hindus. Today, over 900 million people associate themselves with the Hindu faith, Hinduism. Hinduism began in the area drained by the Indus River in India, and the word Hindu was first used by Arab invaders to refer to people who lived across the Indus River. Sometime between 1500 and 500 BCE, the Vedic religion began and the sacred texts of the Vedas were written. As Wikipedia notes: "The Vedas center on worship of deities such as Indra, Varuna and Agni, and on the Soma ritual. Fire-sacrifices, called yajña were performed, and Vedic mantras chanted but no temples or idols are known. The oldest Vedic traditions exhibit strong similarities to the pre-Zoroastrian Proto-Indo-Iranian religion and other Indo-European religions."

Modern Hinduism emerged from these ancient Vedic traditions and the major Sanskrit epics, the Ramayana and Mahabharata, which were compiled between the late centuries BCE and the early centuries CE. Hinduism is a very open and tolerant religion which is also one of the most complex of all religions, and is essentially related to the union of reason and intuition.

Most Hindus believe that the spirit or soul (called the ātman), is eternal. According to the monistic/pantheistic theologies of Hinduism, Atman is ultimately indistinct from Brahman, the supreme spirit. The goal of life, according to the Advaita School, is to realize that one's ātman is identical to Brahman, the supreme soul. The Upanishads state that whoever becomes fully aware of the ātman as the innermost core of one's own self realizes an identity with Brahman and thereby reaches moksha (liberation or freedom).

In whatever way a Hindu defines the goal of life, there are several methods (or yogas) taught for reaching that goal. The Bhagavad Gita, the Yoga Sutras, the Hatha Yoga Pradipika, and, the Upanishads define several paths that one can follow to achieve the spiritual goal of life (Moksha, Samadhi or Nirvana), including:

- Bhakti Yoga (the path of love and devotion)
- Karma Yoga (the path of right action)
- Rāja Yoga (the path of meditation)
- Jñāna Yoga (the path of wisdom).

This wonderfully peaceful religion does not formally seek to convert anyone to Hinduism, although it does accept converts worldwide. In stark contrast to the other two major world religions, very few people indeed have been killed in the name of Hinduism.

It Only Hurts When I Laugh: Abraham, Jesus and Muhammad

"When I first said that I wanted us to put together a late-night comedy writing team that would only be 80 percent Ivy League-educated Jews, people thought I was crazy. They said you need 90, 95 percent. But we proved 'em wrong."

Jon Stewart—2005 Emmy Award acceptance speech for his incredibly popular "Daily Show"

54% of the world's population consider themselves followers of the Abrahamic religions[viii]. Judaism, Christianity and Islam are all considered to "Abrahamic faiths", which means that, at the very least, they all honour Old Testament Abraham as one of the founders of their religion. Not only that, but all three are monotheistic, conceive of God as a transcendent Creator-figure and source of moral law, while the sacred narratives of all three feature many of the same prophets, histories and places, with minor variations on their roles and the meanings attached to the events depicted. So what is it about human nature that continually focuses on their differences, indeed exploits their differences, rather than transcending them to find their common ground?

Hence, while Buddhism and Hinduism may not be warlike religions, this cannot be said of either Christianity or Islam. Like two boxers in a ring, these two religions have circled each other warily since the formation of Christianity in the centuries after the death of Christ in 32 CE, and the emergence of Islam after the death of the prophet Muhammad in 632 CE. Muhammad is considered by Muslims to be a messenger and prophet of God, the last in a series of Islamic prophets and restorer of the original monotheistic faith (Islam) of Adam, Noah, Abraham, Moses and Jesus.

Muhammad was also a man of action. Having been orphaned early in childhood, he became a trader and travelled extensively, meeting many Christians and mystics on the way. Toward the end of his life he had played a wide variety of roles including diplomat, merchant, philosopher, orator, legislator, reformer, and military general. He had a total of 13 wives and fathered three sons and four daughters. In his final years he conquered Mecca and all of Arabia, and died of a high fever aged 63. Today he lies buried under the Green Dome in Medina.

Muslims believe that the verses of the Qur'an were revealed to Muhammad by God through the archangel Gabriel, on several occasions between 610 and his death. The Qur'an was reportedly transcribed by Muhammad's companions while he lived, although the prime method of transmission was oral. The Qur'an is considered as the key source of Islamic principles and values and is divided into 114 chapters which combined, contain 6,236 verses. The earlier chapters, revealed at Mecca, are primarily concerned with ethical and spiritual topics, while the later chapters discuss social and moral issues.

Muslims pray five times a day to toward Mecca, and prayers consist of verses from the Qur'an chosen to express gratitude to and worship of God. Most devout Muslims also fast to be nearer God, and some manage to make the pilgrimage to Mecca called the "Hajj", at least once in their lifetime. Islamic law (known as the "Sharia", literally meaning: "The path leading to the watering place"), constitutes a system of duties that all Muslims must follow by virtue of their beliefs. Sharia is comprehensive, covering all aspects of life including governance, foreign relations, and punishments for five specific crimes: unlawful intercourse, false accusation of unlawful intercourse, consumption of alcohol, theft, and highway robbery.

As George W Bush discovered a little late in his Presidency, there are two different kinds of Islamic faith: Sunni (80-90 percent of Muslims), and Shia (10-20 percent of Muslims). Shia and Sunni divisions date back to the death of Muhammad, and the question of who was to take over leadership of the Muslim nation. Sunni Muslims agree that the new leader should have been elected from among those capable of the job. This is what was done, and the Prophet Muhammad's close friend and advisor, Abu Bakr, became the first Caliph of the Islamic nation. The word "Sunni" in Arabic comes from a word meaning "one who follows the traditions of the Prophet."

On the other hand, Shia Muslims believe that following Muhammad's death, leadership should have passed directly to his son-in-law, Ali. Throughout history, Shia Muslims have not recognized the authority of elected Muslim leaders, choosing instead to follow a line of Imams which they believe have been appointed by the Prophet Muhammad or God Himself. (The word "Shia" in Arabic means a group or supportive party of people).

Most Muslim-majority countries recognize Islam as their state religion. Proselytism on behalf of other religions is often illegal. So, in Shia Iran and parts of Iraq, your Sunni views will get you into trouble. In Sunni Afghanistan, Algeria, Bangladesh, Brunei, Comoros, Egypt, Aceh Province of Indonesia, Jordan, Libya, Maldives, Malaysia, Mauritania, Morocco, Pakistan, Qatar, Saudi Arabia, Somalia, Tunisia and the United Arab Emirates, you will be arrested for trying to convert anyone to Christianity, and not a little unwelcome if you attempt to promote your Shiite, Buddhist, Shintoist, Hindu, or other religious views. Sufis and Bahai believers appear to be tolerated in many of these places provided they keep a low profile.

This then brings us neatly to the Christians, who share (mostly unwittingly), so many important ideas, principles and prophets with both the Jewish and Islamic religions (and most others, when you really get down to their orienting generalizations and advice). Christianity is the world's largest religion with over 2 billion adherents. In some countries it is still the official state religion. For example, the Queen remains the head of the Anglican Church in England, and in Denmark, Iceland, Norway and Finland the Lutheran church is the state religion. If you go to Georgia, you will find the Georgian Orthodox church tells people how it is.

To paraphrase Somerset Maugham, I myself am "of the faith" in the sense that that the church I currently do not attend is Methodist, which is where I was christened. I also spent a fair bit of time in the pews of Lutheran (Bach's organ music was a blast) and Baptist churches in my youth, and was most impressed by the Canadian Unitarians as a youngster. (My parents were sophisticated enough to try to find somewhere to go together on a Sunday morning which appealed to my mother's Christian sensibilities and my father's pantheistic, philosophical approach to religion).

During my life I've attended services in synagogues, lit candles in Catholic churches, visited several beautiful mosques, and joined Buddhist sanghas. But by the age of 21, I had worked out as logically, scientifically and independently as any self-assured 21 year old can, that there was definitely not a God in the singular—or even any gods, plural. There was just the universe, resplendent in all its glories, and effectively infinite in its reach both in time and space. I was the perfect objective, modernist intellectual, expecting science to ultimately reveal all.

My faith was in science and human reason. The alternatives of thousands of years of religious warfare, toxic gurus/religious leaders and animistic superstitions did not appeal to my peace-loving, sharp young mind. What human beings need is to grow, develop and make progress, I reasoned. Or as my grandmother always used to say: "God helps those who help themselves"—and she was definitely not talking about shoplifting.

Having grown up in a liberal, broadminded, "power of positive thinking" Christian family with a dash of Jewish blood on my mother's side, I took it for granted that people in general were inclined to be good and see the best in every situation and others. Early on in my churchgoing days I took the lesson of "turning the other cheek" to heart, and spent months letting the bullies at Woburn Public School in Toronto Canada, punch me without retaliating. I often returned home with a bloody nose and bruises, as I had the misfortune of having been accelerated two years ahead of my age group due to being "gifted". The regular gift I got for being gifted and coming first in everything was to be singled out for bullying by some of the "slower", but very strong, big kids.

My mother was sympathetic, and cleaned my wounds up on a regular basis, but my father was furious. I will never forget the day one of the bullies stole my younger brother's football, and I asked him to give it back to him. After taunting me as to "What, exactly I was going to do about it?" I confronted him with my righteous rage, and challenged him to a fist fight. Luckily I had taken boxing lessons at primary school and had developed a knock-out right cross with a swift left upper-cut. Sensing big trouble, my brother ran home and returned with my father (in retrospect, rather hilariously, riding on my brother's bicycle with a "Banana Saddle" and "Monkey Bars" (think "Easy Rider"), and told me to "finish it off like a man". I did.

I swaggered into school the next day a hero, as word had got around about this earth-shattering event—it was David vs Goliath, all over again. The bully became one of my best friends, and I still have the four 1966 Canadian silver dollars he gave to me as a leaving present when my family moved back to South Africa in 1968. The gifted "chicken" had triumphed.

That, then, was the end of my "turning the other cheek" days. I also began to develop doubts about what I was being taught at Sunday school. If God, who I prayed to regularly, was on the side of the good guys, then why did good guys lose so often? I had discovered what philosophers call "The Problem of Evil". Later on in life, I would add to my list of reasons not to believe in an all-powerful Deity, the "Why do Bad Things Happen to Good People?" dilemma.

Dr Robin Lincoln Wood

Of Christians, Emperors, Bibles and Gypsies

Christianity (from the Ancient Greek word Χριστός, Khristos, "Christ", literally "anointed one") is a monotheistic religion based on the life and teachings of Jesus as presented in the gospels and other New Testament writings. Christianity teaches that Jesus is the Son of God, God having become human and the savior of humanity.

Christianity began as a Jewish sect in the mid-1st century. Originating in the eastern Mediterranean coast of the Middle East (modern Israel and Palestine), it quickly spread to Syria, Mesopotamia, Asia Minor and Egypt, rapidly growing in size and influence, and by the 4th century had become the dominant religion within the Roman Empire.

After solidifying his position to gain complete control of the western portion of the empire in 312, the Emperor Constantine instituted the Edict of Milan, a "Magna Carta of religious liberty," which eventually changed the Empire's religion and put Christianity on an equal footing with paganism. Almost overnight the position of the Christian Church was reversed from persecuted to legal and accepted. Constantine began to rely on the church for support, and it on him for protection. The Church and the Empire formed an alliance, which remains to this day. Very rapidly, the laws and policies of the Empire and the doctrine of the Church became one with Constantine as the interpreter of both law and policy. This was accomplished by eliminating hundreds of books thought to be against "Church" doctrine and watering down what remained by blending Christian beliefs and practice with long established Roman sanctioned pagan worship.

Constantine worked hard to ensure that Church and the State were integrated as much as possible. While on the one hand he tolerated pagan practices, keeping pagan gods on coins and retaining his pagan high priest title "Pontifex Maximus" in order to maintain his popularity, on the other he began a subtle assault on paganism by combining pagan rituals with Christianity. He made December 25th, originally the birthday of the pagan Unconquered Sun god, the official holiday now celebrated as the birthday of Jesus. He also replaced the weekly day of worship by making rest on Saturday unlawful and forcing the new religion to honour the first, not the seventh day, as a day of rest.

As a way of defining his concept of the new universal religion he simply classified everything "Jewish" to be an abomination. Considering that almost every aspect of the Bible is "Jewish" by association, this did not leave much that pleased the Roman Emperor. After 337 Constantine increased his purging of the more obvious aspects of paganism.

Through a series of Universal Councils, he and his successors created their own version of religious doctrine, set up a church hierarchy of his own design, and established a set of beliefs and practices, which today remain the basis for all mainstream Bible-based churches. Very little has changed in Christian doctrine since the 4th century Councils changed the face of Christianity. Over 80% of the total number of biblical books and scriptures that did not meet the new doctrine were purged. The earlier doctrines and practices remaining in the surviving books were effectively replaced with Church-sanctioned doctrine.

In A.D. 325 Constantine convened the First Council of Nicaea in Bithynia (present-day İznik in Turkey) to attain consensus in the church on some of the more contentious issues that threatened to split the Roman Empire in two. Over 300 bishops responded to the invitation sent out by Constantine to the 1 800 bishops of the then Christian church in the Roman Empire, and they each brought with them two priests and three deacons, so the total number of attendees was probably north of 1 800.

The main accomplishment of the First Council of Nicaea was the settlement of the big issue of the relationship of Jesus to God the Father. The council affirmed what it believed to be the teachings of the Apostles regarding who Christ is: that He is the one true God with the Father. They also conducted some other business on their crowded agenda, including drafting the first part of the Nicene Creed; settling the calculation of the date of Easter; and promulgating early canon law.

Having settled that, and narrowly avoiding a split between the folks in the East who believed Jesus was simply one of a long line of messengers of God, and the folks in the west who believed he was actually God himself, things continued pretty much as usual for another hundred years or so. But, as with any religion (or any other knowledge, for that matter), this ancient, it gets more confusing. Two thousand years is a long time, long enough for a myriad of different interpretations to emerge around who said what, and who can tell us all how it really is.

For example, the "canonical Bible", with Old and New Testaments, the kind that Gideon leaves in your hotel room[ix], is the result of the deal done at Nicaea in 325 CE. 300 male bishops chose 73 books for the Catholic version of the Bible, all written by men, of course. Unsurprisingly, there is not universal agreement on whether the books that were selected to be included in the official version of the Bible as approved by the Holy Roman Empire, were actually the whole truth and nothing but the truth. The books of the Bible that are considered canonical number 24 for Jews, 66 for Protestants, 73 for Catholics, and 78 for most Orthodox Christians.

The Gnostic Gospels are works reflecting the Gnostic take on Christianity, which focuses more on the authority of our inner knowledge rather than a literal reading of the books deemed acceptable by Constantine and his 300 bishops. According to the Gnostic tradition, the answers to spiritual questions are to be found within, not without. Furthermore, the gnostic path does not require the intermediation of a church for salvation, and resembles very closely Eastern approaches to enlightenment, reinforcing the probability that Jesus and some of his disciples were influenced by these teachings and possibly even visited some of their authors.

The Gnostic Gospels are a collection of about fifty-two texts based upon the ancient wisdom teachings of several prophets and spiritual leaders including Jesus, written from the 2nd to the 4th century CE. Some have been known for centuries, but previously unknown works, the Nag Hammadi scrolls, were discovered in Egypt in 1945.

Some modern scholars and religious writers have seized upon various passages from the Gnostic Gospels as indicative of a competing, woman-centered element of early Christianity, especially a passage from The Gospel of Mary in which Jesus kisses Mary and the apostles express envy of His love for her.

Gnosticism was rejected by Christianity, but not because of gender issues. Its claims (two gods, a belief that the created world was evil) were simply inconsistent with the rule of faith, as it was called, handed down from the apostles. The canonical Gospels all date from the middle to late first century. Gnosticism was a dualistic, esoteric mode of thinking that was widespread during the early Christian era, although its influence was not confined to Christianity.

The Trouble with Paradise

Dan Brown's novel, the "Da Vinci Code"[x] refers to cryptic messages supposedly incorporated by Leonardo Da Vinci in his artwork. According to the novel, Leonardo was a member of the ancient secret society, the "Priory of Sion", which was dedicated to preserving the truths that Jesus designated Mary Magdalene as His successor, that His message was about the celebration of the sacred feminine, that Jesus and Mary Magdalene were married and had children and that the Holy Grail of legend and lore is really Mary Magdalene, the "sacred feminine," the vessel who carried Jesus' children.

Here in Catalonia, just north of Perpignan, lies the small seaside town of Sainte Marie de la Mer. Every year, a festival is held celebrating the arrival of Mary Magdalene and her children on a small boat, with villagers and tourists alike taking part, running into the sea and greeting the boat used to represent the event celebrated by villagers for perhaps nearly two thousand years.

Sainte Marie de la Mer and Perpignan are among the many places the gypsies began their journey in Europe. Today Perpignan has one of the largest populations of "Gitanes" (yes, like the cigarettes), in France, which have been happily settled here for over a millennium[xi]. In the Arab and Gitane quarters of pre-medieval Perpignan, one can still see the Indian-looking descendants of the original settlers. The Gitanes are also intensely Christian, often favoring the Pentecostal versions of Christianity.

There are many theories about what happened in the 18 years of Jesus' life between the age of twelve and thirty, when he began his ministry. It is really quite incredible that a religion as large as Christianity can be based on twelve years in the life of an enlightened being who lived for 32 years. So, what is being left out, and how reliable are the accounts of this missing time?

The gospels have accounts of events surrounding Jesus' birth, and the subsequent flight into Egypt to escape the wrath of Herod, followed by the settlement of Joseph and Mary, along with the young Jesus in Nazareth. There also is that isolated account of Joseph, Mary, and Jesus' visit to the city of Jerusalem to celebrate the Passover, when Jesus was twelve years old.

Other than the generic allusion that Jesus advanced in wisdom, stature, and in favour with God and man, there is a giant blank between Jesus' twelfth and thirtieth birthdays in both the Catholic and Protestant

Bibles. A common assumption amongst Christians is that Jesus simply lived in Nazareth during that period, but there are various accounts that present other scenarios, including travels to India.[xii]

Several authors have claimed to have found proof of the existence of manuscripts in India and Tibet that support the belief that Christ was in India during this time in his life. Others cite legends in a number of places in the region that Jesus passed that way in ancient times. The Jesus in India manuscript was first reported in modern times by Nicolas Notovitch in 1894. Subsequently several other authors have written on the subject, including the religious leader Mirza Ghulam Ahmad (founder of the Ahmadiyya movement in Islam) in 1899, Levi H. Dowling (1908), Swami Abhedananda (1922), Nicholas Roerich (1923-1928), Mathilde Ludendorff (1930), Elizabeth Clare Prophet (founder of Ascended Master Teachings New Age group) in 1956, and more recently Holger Kersten in his book Jesus Lived in India, published in 1981.

What is certain is that none of the established brands of Christianity have anything to say about such possibilities, apart from bland denials that anyone else could possibly know more than they do about the life of one of the world's most significant historical figures. Deepak Chopra's book: "Jesus", is both an inspirational and well-informed attempt to tell the full story in a way that will appeal to all spiritual seekers as well as normally dogmatic Christians, Islamists and those of other religions who are open to learning new things.

King Size Cash Flow Crises, Crusades and Knights Templar

"We are placed in this world, as in a great theatre, where the sources and the causes of every event are entirely concealed from us; nor have we sufficient wisdom to foresee, or power to prevent those ills, with which we are continually threatened. We hang in perpetual suspense between life and death, health and sickness, plenty and want; which are distributed among the human species by secret and unknown causes, whose operation is often unexpected, and always unaccountable.

These unknown causes then become the constant object of our hopes and fears; and while the passions are kept in perpetual alarm by an anxious expectation of the events, the imagination is equally employed in forming ideas of these powers, on which we have so entire a dependence."

David Hume. Four Dissertations (London, 1757) p172

Of course, any discussion about Christianity would be incomplete without a mention of the Crusades, which began, as with many of the major events in the life of the Christian religion, with a King, a Pope and a severe cash flow problem. The King was Philippe the Fair, aka Philippe the Fourth, King of France from 1285 until his death in 1314. The Pope was Clement the Fifth, who reigned briefly over Christendom from 1305 until his death in 1314. Clement was elected as a compromise candidate to settle a dispute between the Italian and French cardinals, both of whom initially refused to accept the opposition's candidate for Pope, for reasons which will become clear in a moment.

Neither Philippe nor Clement were held in very high regard by their contemporaries. Philippe became King at 17, and the kindest thing that could be said about him and repeated in mixed company was that he was a "useless owl", and that "He is neither man nor beast. He is a statue". The cash flow crisis began when the English conquered Philippe's French armies to win Gascony back. (Although the French eventually took this back during the subsequent 100 years' war, today the British have recolonized it by retiring there and buying large numbers of holiday homes in the Dordogne region, thereby resulting in it being christened: "Dordogneshire").

In the short term, Philippe arrested Jews and confiscated their assets to plug the yawning chasm in his finances. When he ran out of Jews, he started on Lombard bankers and abbots. His next move was to debase the coinage, and then tax the clergy on half their income, though this provoked an ongoing battle with the Pope who preceded Clement, Boniface VIII, leading to the big dispute between the Italians and the French which led to Clement, a Frenchman who had done a secret deal with Philippe, being elected Pope. (One might comment using that perennial French phrase "plus ca change, plus c'est la meme chose", on how little has really changed in politics and religion in the last thousand years. The only major difference appears to be the scale on which leaders of today's massive nation states can debase their coinage and tax everything in sight).

Now that Philippe had his own man sitting in the Vatican-(it was now Avignon in southern France where the new Pope lay his head every night), he could plug the sieve-like financial situation his lavish spending habits created. The next logical target for a shakedown was the richest order of all time: the Knights Templars[xiii].

Philippe was hugely in debt to the Knights Templar, a monastic military order who had been acting as bankers for some two hundred years. As the popularity of the Crusades had decreased, support for the Order had waned, and Philippe used a disgruntled complaint against the Order as an excuse to disband the entire organization, so as to free himself from his debts. On Friday, 13 October 1307, hundreds of Knights Templar in France were simultaneously arrested by agents of Philippe the Fair, to be later tortured into admitting heresy in the Order. If you are superstitious about Friday the 13th, now you will not only know why this combination of day and date is so significant, but you will also get to keep your head, unlike the unfortunate Templar Grand Masters.

The Knights Templar were supposedly answerable only to the Pope, but Philippe used his influence over Clement V, who was largely his pawn, to disband the organisation. Clement did attempt to hold proper trials, but Philippe used the previously forced confessions to have many Templars burned at the stake before they could mount a proper defense.

In 1314, Philippe had the last Master of the Templars, Jacques de Molay burned at the stake. The account goes as follows: "The cardinals dallied with their duty until March 1314, when, on a scaffold in front of Notre Dame, Jacques de Molay, Templar Grand Master, Geoffroi de Charney, Master of Normandy, Ilugues de Peraud, Visitor of France, and Godefroi de Gonneville, Master of Aquitaine, were brought forth from the jail in which for nearly seven years they had lain, to receive the sentence agreed upon by the cardinals, in conjunction with the Archbishop of Sens and some other prelates whom they had called in. Considering the offences which the culprits had confessed and confirmed, the penance imposed was in accordance with rule—that of perpetual imprisonment.

The Trouble with Paradise

The affair was supposed to be concluded when, to the dismay of the prelates and wonderment of the assembled crowd, de Molay and Geoffroi de Charney arose. They had been guilty, they said, not of the crimes imputed to them, but of basely betraying their Order to save their own lives. It was pure and holy; the charges were fictitious and the confessions false. Hastily the cardinals delivered them to the Prevot of Paris, and retired to deliberate on this unexpected contingency, but they were saved all trouble.

When the news was conveyed to Philippe, he was furious. Only a short consultation with his council was required. The canons pronounced that a relapsed heretic was to be burned without a hearing; the facts were obvious and no formal judgment by the papal commission need be waited for. That same day, by sunset, a stake was erected on a small island in the Seine, the Isle des Juifs, near the palace garden. There de Molay and de Charney were slowly burned to death, refusing all offers of pardon for retraction, and bearing their torment with a composure which won for them the reputation of martyrs among the people, who reverently collected their ashes as relics'.

According to legend, de Molay cursed Philippe from the flames, saying that he would summon him before God's Tribunal within a year; as it turned out, he died within the next year. The throne passed rapidly through Philippe's sons, who also died relatively young, and without producing male heirs. By 1328, his line was extinguished, and the throne had passed to the House of Valois.

I am currently sitting writing this sentence about three meters below a two-story Knights Templar Chapel which we discovered upon renovating our derelict Chateau in Perpignan in 2004. The stained glass window in the chapel represents the lamb and the golden sword of the Templars, identical to another window in the Temple Church in London [xiv]. The successive Barons Despres who lived here since the early 1880's and remodeled the Chateau[xv] in the Belle Époque along Italian Gothic lines with a dash of Art Nouveau here and there, clearly worshipped in this chapel with their families and staff for more than a century.

Although links to the Templars abound in this region, with "Celliers des Templiers" wines, "Route des Templiers" in Perpignan and along the coast, and an ancient Commanderie between Perpignan and Thuir, (a small village and ancient Roman fort several kilometers to the west toward the Pyrenees), we have not been able to find out much about the

Baron Despres' links to the Templars, though it is clear that the Despres family was powerful and influential in the region.

During the Middle Ages, most of the remainder of Europe was Christianized, with Christians also being a large religious minority in the Middle East, North Africa, Ethiopia and parts of India. Today, the three largest Christian brands in the world are the Roman Catholic Church, the Eastern Orthodox churches, and various forms of Protestantism. The Roman Catholic and Eastern Orthodox patriarchates split from one another in the East-West Schism of 1054 AD, and Protestantism came into existence during the Protestant Reformation of the 16th century, splitting from the Roman Catholic Church.

All Christians basically assert that if you believe in Jesus, you will be saved from death, Hell and having a sinful character, through a process of positive moral change. For Catholics, there has always been a great deal of ritual involved in this process, administered by the priesthood. Eastern Christianity, on the other hand, views salvation more as a process of healing rather than the more legalistic Catholic approach of grace and punishment. The Orthodox Church supposes that Christians are saved not only by their good deeds, but also by through patient suffering of grief, illness, misfortune and failure.

Protestants, however, decided to "get back to the basics" of earlier Christianity, after a millennium of increasing corruption and worldliness in the church. By the time the fund-raising campaign by Pope Leo X to finance the renovation of St. Peter's in Rome began, the rot had really set in hard. Money was tight, and the clergy had to find a way to raise cash fast.

In the north of Germany, the good Archbishop of Mainz had borrowed heavily to pay for his high church rank and was deeply in debt. He agreed to allow the sale of the indulgences in his territory in exchange for a cut of the proceeds. Essentially, an indulgence could be bought from the Church absolving you of all your sins and ensuring your place in heaven. Given the convincingly bad press Hell has always been given by the Christians, this became a very popular practice.

Although many thousands of Bibles had already been printed on Gutenberg's new-fangled printing press, this world changing machine truly found its "killer app" in the issuing of indulgences by the Catholic Church in the late 1400's and early 1500's, as greedy clerics profited enormously by printing indulgences on a massive scale[xvi] to finance their ongoing real estate development projects and finance the costs of the elaborate banquets and fine robes they required to be of service to their God and impress the local yokels.

And then along came Martin Luther, a hardworking, honest and highly intelligent German priest and professor of theology. Luther was initially not aware of what the good Archbishop of Mainz was up to. Even though Luther's prince, Frederick III, and the prince of the neighboring territory, George, Duke of Saxony, forbade the sale of indulgences in their respective lands, Luther's parishioners traveled to purchase them. When these people came to confession, they presented their plenary indulgences which they had paid good money for, claiming they no longer had to repent of their sins, since the document promised to forgive all their sins.

Luther was outraged that they had paid money for what was theirs by right as a "free gift from God", and felt compelled to expose the fraud. This exposure was to take place in the form of a public scholarly debate at the University of Wittenberg. The Ninety-Five Theses outlined the items to be discussed and issued the challenge to any and all comers.

It used to be that most schoolchildren were taught that Luther very dramatically nailed his 95 theses to the front doors of the Castle Church in Wittenberg, Germany. In reality, Luther presented a hand-written copy, accompanied with honorable comments to the archbishop Albrecht of Mainz and Magdenburg, responsible for the practice of the indulgence sales, and to the bishop of Brandenburg, the superior of Luther.

It wasn't until January 1518 that friends of Luther translated the 95 Theses from Latin into German, printed, and widely copied, making the controversy one of the first in history to be aided by the printing press. Within two weeks, copies of the theses had spread throughout Germany; within two months throughout Europe. Luther's writings circulated widely, reaching France, England, and Italy as early as 1519.

Luther's efforts to reform the theology and practice of the church launched the Protestant Reformation. Beginning with the 95 Theses, Luther's writings spread the ideas of the Reformation throughout Europe. A single Renaissance printing press could produce 3,600 pages per workday, compared to forty by typographic hand-printing and a few by hand-copying. Books of bestselling authors like Luther or Erasmus were sold by the hundreds of thousands in their lifetime.

This bible is the best-selling book of all time—it is estimated that since the printing press began churning them out around 1500 or so, several billion copies have been "sold"[xvii], although how many of those are actually read is harder to estimate.

The core ideas and teachings of Christianity have certainly played a positive role in the emergence of western civilization, from two perspectives:

First, Christianity spread the idea of the "golden rule", an ancient idea first mentioned by Confucius around 500 BCE and replicated in ancient Egypt, Taoism, Buddhism, Zoroastrianism, Sufism, Islam and the rest of the world's major religions. The golden rule simply states that one should treat others as one would like others to treat oneself, which is the foundation of ethics and human rights. This ethic of reciprocity was present in various forms in the philosophies of ancient Babylon, Egypt, Persia, India, Greece, Judea, and China.

Secondly, the Christian doctrine of forgiveness of one's enemies led to the possibility of making peace, love not war, though that has taken a full two thousand years to begin to get a foothold in our consciousness and behavior as a species. Despite the many ills and evils of the established churches prior to the Renaissance, followed by the often unintended negative consequences of missionaries and their ilk after the Protestant Reformation, there has also been a great shift in the morality of our western world thanks to Christians who walk their talk, and model the change they wish to see in the world, just as Gandhi reminded us from halfway around the world not so long ago.

Following the Age of Discovery and then accelerated by the Protestant Reformation, Christianity spread to the Americas, Australasia, sub Saharan Africa and the rest of the world through missionaries and colonization. In order to follow Jesus' command to serve others, Christians established hospitals, churches, schools, charities, orphanages, homeless shelters, and universities in the areas to which they emigrated.

At the same time, the persecution of the French Huguenots by Louis XIV and XV forced them to flee Europe to the New World countries, and the rise of the British Empire was tempting either desperate or upwardly mobile English, Irish and Scottish citizens to emigrate to the New World. One of the most promising, but distant of these New World destinations lay at the southern tip of Africa.

The Trouble with Paradise

Beginnings

"Ex Africa semper aliquid novi"—"Always something new from Africa"

Pliny the Elder, Naturalis Historia, VIII/42

"Being from Africa is the best thing that could have ever, ever happened to me. I cannot see it any other way. All of my fundamental principles that were instilled in me in my home, from my childhood, are still with me."

Hakeem Olajuwon-Muslim American Basketball Star

I was born in Africa. Along the shores where the Indian Ocean meets southern Africa, lies a steamy, buzzing port, Durban, full of the spices and promises of the East, yet reeking of the dark heart of Africa. The pungent aroma of strong curry, ripe armpit and exotic perfume mingles in the gentle breezes with the smells one would expect from a large metropolis, with a special mention for the pungent aromas of warm earth after a sudden shower, and floor wax polished to a mirror finish by Zulu elbow grease, then baked under the blazing sub-tropical sun.

Past the harbor, the yacht basin and the millionaire's mile of expensive hotels, runs an expressway carrying thousands of people an hour past the throbbing heart of the city. Along this expressway, standing like sentries guarding the land from an invasion from the sea, runs an unbroken alley of tall palm trees, their leaves glistening in the midday sun. This highway eventually wends its way through the hills of Zululand, the pastures of Natal and the plains of the high veldt to the city of gold, Johannesburg.

If you stand in the hills of Zululand and look up at the night sky, you will see a trillion stars. In a sky so deep you feel you will fall off the earth into its infinite depths, the Milky Way galaxy stretches out before your eyes like an infinite web of translucent cream woven into the fabric of the velvet night. As you stare into the heart of the galaxy from its outer edges, from a small blue planet orbiting a small yellow sun, your total insignificance becomes clear. Fourteen or so billion years on from a big bang, you, a tiny blob of protoplasm, are now the universe, aware of itself.

Yet, there was always something more—a deep longing in my heart, as if my home was somewhere out there, amongst those dense stellar clusters, and I, a foreigner on this fragile oasis in space. I would ask myself many questions: Why does earth seem such a strange place? Why does so little make sense here? Why are we here? And where are we going? To these big questions, I found few answers. But I knew for sure that if I had to land anywhere on earth at any time, then the closest thing I could find to paradise in the galaxy would have been Durban in 1956.

I grew up in two totally different worlds, which created both confusion and entertainment in my youth. On the one hand, my family came from opposite sides of the divide between white South Africans. My father's family was predominantly of Anglo-Saxon 1820's settler stock who had grown up speaking Xhosa and Zulu, while my mother's family was predominantly Afrikaans (descended from the original Dutch settlers), with liberal sprinklings of French Huguenot, and Lithuanian immigrants spiced up with the odd Jewish settler here and there.

Between the 1750's (when my mother's original theologian ancestor left the freezing cold shores of northern Holland to build the first church in Stellenbosch and a beautiful French colonial home called 'La Gratitude'), and the early 1960's, (when my family emigrated to Canada), life had changed a great deal for everyone in South Africa. In those two hundred or so years, people of colour had come to the conclusion that they had not been treated entirely fairly in those first few centuries by whomever was in charge, whether English or Afrikaans.

As a result of the Sharpeville riots and an organization called the African National Congress led by a brave young lawyer named Nelson Mandela, we emigrated to Canada in early 1963, where my parents no doubt thought we would be safer. They did not know then of the dangers of drugs, rock and roll, and even transcendental meditation, being exported by musical groups with long hair such as the Beatles, which were about to hit the shores of Lake Ontario with a vengeance. Canada appeared to be a safe distance away from Xhosa pangas (a kind of machete wielded by some of the more extreme terrorists, who would hack unsuspecting 'settlers' to death) and Zulu assegai's (spears used by the Zulus), so Toronto became our new home.

The Trouble with Paradise

After an early youth in one of the most conservative outposts of the British Empire, I ended up in Toronto, Canada, probably one of the most liberal places on planet earth in the 1960's. Canadian culture and values in this decade were pretty much at the forefront of what was happening planet-wide, especially in the form of one Professor Marshall McLuhan. When he wrote *Understanding Media*, McLuhan was Director of the Center for Culture and Technology at the University of Toronto, which investigated the "psychic and social consequences of technological media." He was not interested in the content carried by a medium, but in the psychic and social effects inherent in the way it extended our senses. As he noted, media like the electric light and electric power grid have no content whatsoever, yet they have significant impact.

Eventually, with the end of the 1960's came the end of our family's love affair with Canada. Being far from the familiar comforts of home and family had left my mother with a serious case of homesickness, and, despite South Africa having left the Commonwealth in 1961, things had apparently improved in the sense that the government had succeeded in arresting most of the communists and "terrorists" (including Nelson Mandela himself), and the country had become a magnet for foreign investment. I missed skiing, ice hockey, the snow, my buddies and the open-minded Canadian culture, but instead said hello to sunshine and monster thunderstorms that would sweep the sky clean every summer's afternoon and evening.

As a teenager at school in Johannesburg, I could not see the relevance of learning French, or French history, in the middle of Africa, where few spoke it. Although I belatedly confess that learning about the French Revolution actually came in handy over the years (as educated people have a habit of taking things like that for granted, you can appear more intelligent when you understand what they are saying rather than grunting: 'Huh?'). And industrial drawing came in really handy, proving to be a truly valuable asset when designing the interiors of the various houses I have renovated, not to mention the gardens and pool area and fountains in the chateau.

And the French lessons eventually came in especially handy in renovating an old Chateau in the south of France, from attempting to describe the kinds of cornices and window frames and different types of plaster needed to enable our architect and builders to do a great job.

But it would take me several years to move beyond the "ordering in a restaurant" French of my youth to something resembling the real thing.

My father started out as an advertising man and copywriter, and it seems every copywriter has a secret novel deep inside him. He used to bang out his memories of his Transkei youth on an old Remington typewriter. The reproductively prolific British side of my family were, inter alia, suppliers and bankers to the royal Mandela tribe. (My father also spent much time on business plans that should have made our family as rich as we supposedly deserved to be, having had all our farms and general stores expropriated by the apartheid-era South African government for the sake of an ethnically "pure Xhosa" homeland in the Eastern Cape.)

This unwieldy typewriter was passed onto me as a youngster, and I too, toyed with the idea of writing my great novel on it for many years. The first computer keyboard I encountered was that of an Apple 2 during my first real job as a young investment banker. With the help of VisiCalc and Basic, I was able to produce complicated financial models that impressed my Vice President boss. I had encountered the new medium of our time, although the definitive South African novel was still eluding me despite several promising outlines.

Excluding newspaper routes, student jobs, and a couple of years as an army officer, I have spent most of my life making a living in one kind of business or another. During those two decades and in the decade of study that preceded them, I found I had to split myself into two very different people in order to cope with the world of business. One part of me (let us call him the artist-philosopher), found the whole notion of business highly unattractive and even questionable. I questioned the greed, fear and machismo which appear to drive a great deal of the world of business.

Business, particularly the version practised in the United States of America, tends to be somewhat "red in tooth and claw", paying little attention to its side-effects and to the "losers". On the other hand, another part of me (let us call him the scientist-historian), found business and in particular economics and technology, to be a fascinating field. Here, if one can remain unemotional and detached, is a way to understand the evolution of our modern world and even perhaps, its future.

My biggest concern over the years has been that our western way of life consistently values money and "stuff" more than it values people.

It does not have to be this way: although it is clear from historical accounts that life was not exactly a bed of roses during our existence as hunter gatherers and farmers, the closeness of the small communities these ways of making a living gave rise to, may have compensated to some extent for the harsh challenges these lifestyles pose. The industrial revolution gave rise to a philosophy in which the machine became the measure of all things. The alienation and social breakdown that resulted have been closely studied and recorded over the past three centuries. Only the rise of the state and large corporations were able to impose order on the scale required to match the social chaos industrialism created.

And yet, as I write these words, this modern order is itself disintegrating in the raging hurricane created by the forces our modern culture and technologies have unleashed around our planet. There are many theories about how we should all deal with the challenges this turbulence, complexity and uncertainty is causing us. Stripped to their essentials, they are variations on themes which have been played repeatedly throughout recorded history, and to which we will now turn.

A Particularly Christian Paradise: The New World

"I am, only because you are."

Bushman wisdom

Like many New World inhabitants, many of my mother's and father's ancestors were escapees from religious intolerance and wars in Europe. My matrilineal ancestor, Meent Borchardus Borcherds was born in Jemgum, East Friesland, in the Netherlands in 1762. He was the only recorded child of Borchardus Borcherds, whose ancestors and wife's identity have disappeared into the mists of history. It is recorded in the memoirs of Pieter Borchardus Borcherds, the eldest son and one of ten children of Meent, that his father left the chilly climes of the northern Netherlands due to an illness. Upon his arrival in the Cape of Good Hope, things were a little tight for the good Reverend for a time, as is evidenced in the archives of the Cape[xviii].

For almost half a century from 1786, Ds Meent Borcherds was the formidable Dutch Reformed Church minister at Stellenbosch. Personally pious, a gifted minor poet and antiquarian, and free of any taint of 'enthusiasm', this East Frisian immigrant was an example of what was best in the 18th century Dutch church, and better yet, from the government's perspective, he was far too rational to succumb to the new impulses of emotional evangelicalism which arose in the last years of that century, and were deemed to be a danger to good order.

Though himself a slave-owner, Borcherds was sympathetic to the aims of the missions to the slaves in Stellenbosch. He regularly performed important roles in ceremonies, such as the introduction of a new missionary or the opening of a new place of worship. He tried to set an example by propagating the Gospel among the slaves of his own household.[xix]

In 1798 Meent Borcherds built one of the most interesting houses remaining in Stellenbosch, called "La Gratitude". In the memoirs of his son, Petrus Borchardus Borcherds, we read that: "He placed in the portico the emblem of the All-seeing Eye, which in this the country of his sojourn had watched over him."

From the Preface of the "Historic Houses of South Africa", we find the legendary South African, General J. C. Smuts[xx], has this to say:

"The old Dutch homesteads of South Africa deserve to be better known than they are. In a country where, as a rule, Nature is everything and Art literally nowhere, our Old Dutch houses form the most notable exception to the rule. The genius of South Africa has shown itself in action—in great deeds, heroic sacrifices, and gifts of leadership—rather than in the domain of Art. Neither in Music nor in Literature nor in Painting nor in Sculpture have we anything yet to compare with the performance of older countries. The one exception is our domestic architecture, and there our production is of a unique character. I believe it was Ruskin who said that the only real contribution to Architecture for the last few centuries has been made by the Dutch in South Africa—or something to that effect.

And the truth of this will be clear to all who have studied the noble houses built at the Cape in the eighteenth and early nineteenth centuries. Since then our taste has been debauched by the commonplace or hideous types introduced from abroad. It is only quite recently that Mr. Herbert Baker has taken us back to the old Cape style, and has popularised its distinctive features in many a beautiful house in most parts of South Africa.

It is evident that this noble architecture could only have arisen in times of comparative quiet and leisure. And of this there must have been plenty in the secluded sun-filled valleys of the Cape in those far-off times. The earliest settlers do not seem to have had the strenuous struggle for existence which marks many new countries and must have had time for the amenities of life. People hurried and urged by violent competition have not the time to consider the artistic effect of their houses or to plan gardens in which to enjoy leisure."[xxi]

By the early 1830's things had improved dramatically in the Cape in comparison with the humble beginnings of the good Reverend and the early days of the Cape Colony[xxii]:

"The following summary of information, supplied by Mr. Borcherds concerning society, &c, in 1803, may be of interest. In the Keizers and Heerengracht (now Darling and Adderley Streets) were the residences of fashionable families. Canals, with sluices, ran along the streets. In seats and bowers, opposite the front door, it was not unusual to see the family enjoying themselves—the gentlemen with their pipes, the ladies by taking tea. Society was usually kept up by evening parties. Small circles of six or eight families were alternately formed and assembled in turn at their houses. Smoking and playing d'ombre or Quadrille, were the favourite amusements of the gentlemen, whilst the ladies in a separate room engaged in fancy and other work.

When the Castle gun fired at nine o'clock, they retired, and were borne home in sedan-chairs. Early rising was customary. Officials, &c, went to office between eight and nine o'clock. From eleven to twelve morning calls were made, when bitters, &c. (Amara and others), were presented. Twelve to one or two was the general dinner time. An hour's repose after this was taken. Afterwards the most respectable and fashionable used to dress for either a drive in the country or to be prepared for evening society. The whole community was like one family. At the top of Government Gardens there was no outlet, but a large square, in which a menagerie was kept. In those days there were few balls, but amateur concerts and plays constituted the principal amusements. A club called the "Harmony" existed in Adderley Street."

P.B Borcherds lived a long and fruitful life, leaving 30 grandchildren and countless great grandchildren, and retired to write his memoirs after having been a civil servant for 56 years. His final position was as the Resident Magistrate and Civil Commissioner for Cape Town and the Cape District, reporting to the Governor and Commander in Chief of Cape Town, Sir George Grey, KCB.[xxiii]

P.B. prefaces his memoirs as follows:

"Having been spared to reach my seventy-first year, and to serve my native country as a civil servant for about fifty-six years, and being father of a family of

```
Children, .................................................. 17
Grand-children, ......................................... 30
Great-grand-children, ............................... 10
Children-in-law by marriage, ..................... 12
Total ......................................................... 69
```

with the prospect of leaving a numerous descendancy, and being as yet favored with the partner of my life by my side since 1806, and enjoying a day of rest (having retired from the public service), and thank God! Still in possession of the faculties of mind and body, I undertake the task of giving to posterity a plain account of my life, intermixed with such historical sketches as I may consider of some interest, partly from memory, partly from information, notes, and documents in my possession.

It will, I am confident, be gratifying to my family and friends and pleasing to myself to recall to mind remembrances of day's long past, and give me opportunity of paying a tribute of gratitude to parents and valuable friends for the many acts of love and kindness and consideration which I have been favored with, and especially of those friends to whom I am indebted for promotion in my public career and many blessings which have attended me through life.

It will afford me opportunity to recall to memory my early time, the companions of my youth, the happy days of home and family circle, and the pleasures enjoyed amongst the villagers of that beautiful little town Stellenbosch, where up to my fifteenth year I spent my early days, free from the care and trouble which attend after-life. It will further bring me back to the period when, under the auspices of Government, I had opportunity of seeing my native country, travelling to assist in exploring

The Trouble with Paradise

the interior beyond the Orange River and to open communication with various tribes of the bordering aborigines.

Moreover, it will lead me through the various occupations and employments of a public servant from 1801 to 1857, and leave room to communicate events and observations during the changes of Government and under the administration of twenty Governors and Lieutenant-Governors, and the public men with whom more or less I had the good fortune to come into contact, and to indicate how far I took a share in those institutions which were created to promote education, literature, and science in this colony.

Whilst planning so extensive a course over so wide a field and extending over a period of upwards of half a century, I feel almost discouraged; for my education was restricted to the limited advantages enjoyed by the colony in my youthful days,—days far less favorable and advantageous than the present, and not comparable with the enlarged scale established for mental improvement enjoyed by the youth of our day.

Indeed, it is only trusting that the good intent and object in view will be kindly considered that induces me to make the present attempt, hoping that He who has so mercifully guided me through life will not withhold the means and ability, so that my offspring may read with satisfaction this plain account of their ancestor, that my countrymen who, through their representatives in Parliament, were pleased to bestow on me a public testimonial and recognition of my past career when advanced years compelled my retirement from service, and my other friends may judge of my proceedings during an interesting course of public and private life; and my young fellow civil servants be taught that a strict adherence to duty continued with zeal and integrity, the colonial service ensures ample reward to those who will persevere and surmount obstacles and who are true and faithful to their task, and that the South African public feel and are capable of appreciating and estimating the public servant whose life has been devoted to their welfare and interest.

P. B. BORCHERDS, Senior, Bellevue, Wynberg, January, 1858"

But, that, as they say, was then, and this is now. What would it mean to be able to sit down at the end of your life and write a 21st century set of memoirs as contented as those P.B. committed to paper in the last years of his happy life? Has the world changed so much that a good, long and happy life has become a distant dream for many?

There was clearly not a shadow of a doubt in P.B.'s mind that he was going to heaven, and that his life had been an amazingly productive and wonderful experience. At that time in history, for someone of "common" origins, having to make their way in the world through their own talents and hard work, this was probably as good as it could get.

P.B.'s father, the very Reverend and formidable Meent Borcherds, wrote a great deal of poetry—in fact he was the second person in the history of South Africa to publish a book of poems. The grandson of P.B. translated this poem by him (written originally in High Dutch sometime around the late 1700's/early 1800's) into English in 1937. At the end of this book, I will reveal a third version of this 200 year old poem, written by me in London in the late 1980's. It really is a poem of great staying power and depth, and I am hoping that one of my descendants will follow this up with a 21st century version, once the dust has settled on the first turbulent decades of this century.

The Lake of Eternity

By Reverend Meent Borchardus Borcherds (circa 1800)

(as translated into English from the original Dutch by his great grandson Pieter Borcherds III in 1937)

Much as the billowing of the ocean's waves, thus roll years and generations.

The one advances, the other succeeds and passes on, in anticipation,

And generations have sunk into the lake of eternity.

We who now live are destined to submit to the same fate,

And you, my dear posterity, shall once find, in the flood of time,

Your stream of life glide towards the fathomless lake.

But in the day of judgement we shall revive;

Time will be over and death shall reign no more.

Oh! May I see you all you on that great eve,

rise favoured by the Lord

And with me approach with joy the Almighty Judge's Throne.

For as we are all destined to pass on, be it soon or later,

I resign you to our Heavenly Father,

And may he grant you grace.

Chapter Three

A Good Life, in the 21st Century?

"And what is good, Phaedrus, and what is not good—need we ask anyone to tell us these things?"

Plato

"The medium or process of our time—electronic technology—is reshaping and restructuring patterns of social interdependence, and every aspect of our personal life. It is forcing us to re-consider and re-evaluate practically everything. Everything is changing . . . every thought, every action and every institution formerly taken for granted.

You, your family, your education, your neighbors, your job, your government, your relation to the others, and they are changing. The suddenness of the leap from hardware to software cannot but produce a period of anarchy and collapse, especially in the developed countries."

Marshall McLuhan—The Global Village

If you can afford this book, you probably enjoy the "luxury" of some free time and a pleasant natural environment in which to stand back from your personal version of the rat-race, just by going for a walk in a park or garden at lunchtime. If you go into a park or green space, take a few deep breaths and clear your mind, it will become obvious that there is a choice we can each make. This choice revolves around our ability to be aware of our context, and our options, at every moment, or to not notice that choice and live an unconscious, unexamined life.

And this ability is available to every member of the human race, every moment, thanks to our unique evolutionary heritage, which gave us all an internal universe, centered within our minds and hearts. How do I feel? What do I think? What do I prefer? What do I want? We have the unique capability as a species to be self-aware, in the moment, of what is going on, and to reflect on the feedback we are getting from these sorts of enquiries to our self.

But this requires the energy, time and motivation to enquire and reflect. And of course, it helps a great deal if we have places, spaces, objects, people and rituals to remind and help us to do this. Some people attend ceremonies (religious or otherwise), others go to gurus or analysts, some write diaries, others read, play sport, walk, tour, mow the lawn. Reflection and awareness are available to us at all times, if we can learn to harness the awesome power of our hearts and minds. This is especially important in business, where so many values and priorities are competing for our attention. And for most of us, the point of awareness, enquiry and reflection is to not only understand our world and become better people, but to take right decisions and right action.

One's values and situation can be potentially limiting factors in enabling us to find the space and time to develop such a reflective capacity. Capitalism and the philosophies associated with it place some serious constraints on our ability to develop this capacity. While industrial and financial capitalism may have generated an incredible amount of economic wealth, they have made it more difficult to step back from the fray and reflect on ourselves, our values and our society. In order for us to do this, we actually have to "leave" the capitalist system, either by going on holiday, taking a sabbatical, or "getting out of the rat race".

My biggest concern over the years has been that our world consistently values money and "stuff" more than it values people. Although it is clear from historical accounts that life was not exactly a bed of roses during our existence as hunter gatherers and farmers, the closeness of the small communities these ways of making a living gave rise to, may have compensated for some of the challenges these lifestyles posed. The industrial revolution gave rise to a philosophy in which the machine became the measure of all things. The alienation and social breakdown that resulted have been closely studied and recorded over the past three centuries. Only the rise of the state and large corporations have been able to impose order on the scale required to match the social chaos industrialism has created.

Now, with the rise of the information worker the dysfunctional patterns of industrialism are being transferred to the world of knowledge workers. 21st century call centers, for example (employing hundreds of millions of people around the world today) apply the logic of the factory to the provision of services. Providing you are prepared to key in the exact digits on your telephone (assuming you already know exactly what you want), then wait for several minutes listening to canned music until you can speak to a human being, you may be lucky enough to have your request granted or your question answered. You will naturally have to be able to remember all your passwords and any other secret codes you may have innocently filled in on some form in the distant mists of time.

Of course, call centers and web sites have enabled us to access services twenty-four hours a day seven days a week from anywhere in the world. This convenience is valuable, but it requires hundreds of millions of active people to sit in front of a glowing screen hooked up to a telephone headset for eight hours a day trying very hard to be cheerful. This may not be in the same league as having to work down a coal mine or in a textile mill for twelve hours a day six days a week from the age of eight, but it can be a pretty unfulfilling existence.

The biggest question our system regularly fails to answer is how to enable every person to realise their full potential and their dreams in their work. The idea of a job and a career are products of the industrial age. Work should not be something we do eight hours a day five days a week so that we can have a life. Work should be what we do to become who we are at our best. There should not be a need to run extensive "work-life balance" programmes to ensure that people actually have a life. The point is that over time a gradually larger number of people should be able to claim that their life and work are going to add up to their "life's work".

But, in order to do this, it might just be useful to consider what we mean by "life". We are all acutely aware that one day we will die. And yet, here we are, alive. This ultimate contradiction is ignored by most of us, most of the time, because there is precious little we can do about it apart from doing things that help us live longer. We don't know what happens when we die, nor if we came from anywhere else before we were born. The mystery remains, no matter how hard we think, pray or meditate.

The ancient split between religion, art and science does not help us here. We are alternately told to not question and simply have faith by religions in those areas where no hard evidence is available; to just express ourselves or know what we like in the arts, where what is good is in the eye of the beholder; or to question all assumptions and look for the evidence by the scientists. As it turns out, these ancient splits say more about the evolution of western civilization than they do about the way the world works. These artificial distinctions between what we can know and what we can intuit; what we want and what is "best" for us and those around us; what is good, and what is not—are they blinding us to something deeper, more profound going on below the surface of our lives?

Twenty Years in an Open Necked Shirt

"Quality is the continuous stimulus which our environment puts upon us to create the world in which we live"

Robert M. Pirsig, Zen and the Art of Motorcycle Maintenance-Vintage, 1974.

Zen and the Art of Motorcycle Maintenance made a big impression on me when I first read it back in the late Seventies. I was living in Johannesburg, training to be a lawyer and one of the student leaders of the campaign to release Nelson Mandela. It was clear to my generation of baby boomers that traditional values and logic had become almost useless in their ability to resolve the complex social problems we were creating around us as we drove modernization and scientific positivism to their logical conclusion. It seemed that there was a giant vacuum at the center of our civilization, which many unfortunate souls regularly fell into and from which they never returned.

Having spent several centuries developing and perfecting science, technology and all the accoutrements of the modern world, western civilization was just beginning to get to grips with the social consequences of technological innovation, globalization, and the complexity and disruption which flowed from these developments. We may have solved, at least for two-thirds of the world, the problems of hunger, shelter and getting a job to look after yourself and perhaps your family in a vaguely decent way—yet the bigger problem was still sitting on the horizon: what does it mean to be fully human and to live "a good life" in the maelstrom of the 21st century, with all its mega challenges?

I think it is safe to say that most people would not imagine themselves living "the good life" in a buttoned down shirt with a tie and jacket, apart from making the odd James Bond like appearance for cocktails in a dinner jacket. Yet, for many between 1950 and 2000, being a "white-collar worker" meant a life of relative security in a not-too demanding job in some kind of large organization. My father was inspected at our doorstep by my mother each morning to ensure his tie was straight, shirt unblemished and his trousers well pressed, before he eagerly went off to work in the world of big business.

I was thus a proud little fellow when, at the age of six, I was spruced up and similarly inspected for my first day at school in a white shirt, red, green and yellow striped tie, a grey flannel blazer and shiny black shoes. A grand career in the corporate world obviously beckoned, and the glory of this world was only slightly tarnished by my lack of understanding of what, exactly, my father did all day—he told me he was, as a manager, mainly "having meetings".

Thus it was that, apart from my wild university days in the glorious, psychedelic 1970's, I was dressed between the hours of 9 am and 6 pm in a suit and tie. My career plans were clear, even if they changed almost as quickly as my career: to get to the top of whatever it was that one could climb to the top of, as soon as possible, so that I too, would be a manager and leader of men (until women's liberation came along in the 1960's, women's roles seemed confined to typing memos and making coffee in these organizations, "Mad Men" style).

After a decade in the corporate world, and having "made it" as a manager, consultant and executive, the mutual realization dawned on myself and the heroic leaders of this world that I was congenitally unsuited to being a corporate animal. Thus it was that, in 1990, at the ripe old age of 33, I started my first business. My first act as "CEO" was to declare a smart casual dress code for all employees (which at that stage was me, myself and I), a version of the dress-down Fridays attire that was to become highly popular by the end of the nineties amongst the suited ones. In the early nineties only wealthy jet-setters, trade unionists, artists, designers and successful yuppie business people and technologists like Steve Jobs, Bill Gates, and Richard Branson could get away with open-necked shirts and casual sweaters at business meetings.

And I have, thankfully, with the exception of funerals and black tie dinners, spent the rest of my life in an open-necked shirt or a turtleneck.

My decades in corporate life taught me that working for a large organisation in the twentieth century was a form of voluntary servitude, designed for the benefits of the senior suits, and not Joe Average or me. And the tie was the ultimate symbol of one's submission to the system—why else did sixties rebels wear jeans and look like Hell's Angels. You could not be truly free, or identify with the working classes, and wear a tie. Of course, the tie is only an outer symbol of a much deeper truth—that organisation men and women are conditioned to accept that the ultimate measure of success in corporate life is the hard bottom-line: how much value have I added or how much profit have I made for the company? And the soft bottom line is not far behind the hard: how well am I conforming to the norms, values and behaviors which one needs to get ahead in this firm?

In the nineties debate raged around the concept of the inclusive corporation. Are shareholders the ultimate frame of reference, or should the wider stakeholder community be a consideration in defining the framework by which success is evaluated in a firm? In other words, is it just shareholder returns that count, or is a broader set of measures required against a broader group of stakeholders, for example local communities, the environment, the government, employees, business partners and so on? Both sides brandished evidence to demonstrate that ethical, inclusive corporations outperformed companies run on pure shareholder value terms.

Today that debate is over: corporate social responsibility is an essential focus for every CEO, and sustainability has become not only a moral imperative, but a business opportunity. The challenge of the 21st century is to find ways of co-creating shared value in a more harmonious, equitable fashion than the growth obsessed capitalism bequeathed to us by the 20th century. But how?

Paradise Regained?

"There is . . . an appearance of breakdown which results from the realization of the new values themselves, because these new values are so often the exact antithesis of the old. In that sense, the new values do represent the ultimate breakdown of the current basis of society, or of the individual's way of life."

Clare W. Graves

Ironically, the "mess" we are in right now on our planet is a result of our success as a species—we have created a state of hyper-abundance for a few billion people, accompanied by a state of relative abundance for 3 billion people and poverty experienced by roughly another 2 billion people. We now have the capacity to keep people alive for twice as long as their expected lifespan one century ago. 500 million people died from small pox in the 20th century, but by 1980 we had eradicated that disease. Polio, measles, mumps and rubella are also on the endangered disease list now and should be eradicated in the next few decades, along with malaria. AIDS will take a little longer, but its spread is now diminishing.

3 billion people in the developing world experience life as a continual challenge for survival in the face of age-old stressors such as hunger, disease, dictators, war or natural disaster. In the developed world, however, the stress experienced by 4 billion of us each day is more to do with the natural ups and downs of life against the backdrop of our success in rolling out new, improved products and services and creating excessive amounts of choice and bureaucracy. We are having to deal with the complexity of new technologies, lifestyles and neighbors, with stresses related to over or under employment, being in the right or wrong career, or relationship, or living in our less than ideal location. All these stresses are relative, relating to possibilities and expectations we have, and which we can choose between. The stress of affluenza is relatively new, and can be devastating for the perfectionists among us.

But we must remember that it is we, ourselves, for the most part, who are generating the stress, rather than a threatening set of life conditions in our immediate world. It appears that it is precisely when things cannot get any worse, when the mess cannot get any bigger, that we can be sure that a new set of coping mechanisms is evolving in response to the challenges of the prevailing life conditions, because that is what evolution does. And in the mega-cities and country retreats of the developed world, one can spot a variety of responses to the stresses of affluenza and its accompanying life conditions.

In his seminal study of our post-modern, media-saturated world, Thomas de Zengotita portrays the technologically advanced, media-saturated West as a world composed of millions of individual "flattered selves," each living in its own insulated "ME World." This epic narcissism is constantly being nourished by media representations in all areas of our lives, from personal websites to advertisements on television.

"Our minds are, as a matter of sheer quantitative fact, stocked with mediated entities," he writes. "Ask yourself: is there anything you do that remains essentially unmediated, anything you don't experience reflexively through some commodified representation of it? Birth? Marriage? Illness? Think of all the movies and memoirs, philosophies and techniques, self-help books, counsellors, programs, presentations, workshops . . . and the fashionable vocabularies generated by those venues, think of how all this conditions your experience."

De Zengotita describes the postmodern individual's quality of being as that of a method actor. In a culture saturated with media performances, where one's life is informed by representations of "life," thereby becoming a subtly self-conscious performance. He uses the image of athletes celebrating a victory on television to make his point:

"There's that same element, that same quality in the way those exhilarated men position themselves in front of each other, or the larger audience and the cameras, beefy faces alight with a peculiar blend of exultation and hostility, tendons bulging in their necks, fists pounding the air or curled tight upward at the ends of crook-dangling arms, bodies thrust forward as if to bulldoze past all compromise, apparently frenzied, apparently berserk, bellowing in tones suggestive of profound vindication, bellowing "Yeaauh! Yeaauh! Yeaauh!" And each "Yeaauh" lifts above the preceding one, as if to reinforce it, but also to comment on it, even to parody it, and suddenly you realize, looking into their eyes, beaming out at friends and neighbors in the stands, you realize that this is also a performance, and a contest, a folk art—and oh-so-self-conscious after all."

We have become, he says, "celebrities all, celebrities at last"—the knowing stars in the self-directed movies of our lives.

McLuhan said the electric media would cause a social and political "implosion," raising "human awareness of responsibility to an intense degree." Television has made us aware of global calamity and injustice, and has allowed mass populations to "know" the same celebrities, and share their ups and downs. The Net is also global, allowing us to find information and people regardless of their location. In the context of the 21st century, we form communities of common interest and meaning, as often as those in common locations.

But interest-based communities are explosive, not implosive. As we spend time with them, we have less time for others. Time spent on the web is also time away from our families, neighbors and colleagues down the hall at work. That changes our sensory and emotional balance, and the very concept of "a person" becomes more abstract and partial.

If only a few people use a technology, it cannot transform them or society. Only when it becomes pervasive and taken for granted do its ultimate impacts emerge. The slate and pencil and paper shifted our education system toward literacy. The blackboard nudged us in the direction of lecture, away from reading. When personal computers, mobile telephones and the web achieve the same penetration levels, the world will have changed dramatically again. In half the world it already has.

Our media shape both our society and us. After the fact, the social consequences are relatively obvious. Television transforms consumption, the automobile reshapes the city, and the clock synchronizes work and many other aspects of our lives. The psychic effects are closer, and therefore more difficult to see. Perhaps television shortens our attention span, the automobile enhances a feeling of isolation, and the clock accelerates our internal sense of time (thereby shortening our subjective lives).

The message of the electric light is like the message of electric power in industry, totally radical, pervasive, and decentralized. Medium centralization has at least four dimensions—centralization of access, capital formation for distribution and for terminal equipment, and intelligence. For example, access to TV is centralized since a few people decide what is broadcast, distribution is centralized since cable companies and broadcast stations pay for the infrastructure, the purchase of terminals (TV sets) is decentralized, and intelligence is centralized since the TV set is just a viewing device. Couch-potato media like clothing, radio, newspapers, magazines, and movies follow similar patterns.

The web, like fax, telephone, automobiles, money, and other media, is decentralized in every sense. Users decide when to send mail, post queries, or establish servers. Since the web grows from the edges, like a Tinker toy model, users and their organisations fund their share of switching and access costs, and funding of PCs and workstations, is also decentralized. Decentralization of the web, combined with falling communication cost and increased terminal (computer) intelligence, points to eventual pervasiveness, and therefore a radical impact

The individual qualities that sort us into social and economic winners and losers change with time. Physical speed and sharp eyesight were once keys to success, and the importance of memory was diminished by the invention of writing and printing. Linear thinking may not be as important tomorrow as it was yesterday. Perhaps naturally cooperative people, channel surfers and compulsive social networkers will be the winners in the networked world? We certainly need to develop a much more profound understanding of how our social and cultural evolution works if we wish to live well, let alone take up residence in our own personal versions of paradise.

The downside of all of this is that we have unwittingly created a new form of servitude which comprises billions of people spending trillions of hours glued to small screens developing repetitive strain injury in clinically austere surroundings, trying to make their living. I have worked in corporate cultures where e-mail creates an almost delusional space where employees will spend hours telling jokes, spreading rumours and "having fun". Today we can do the same thing on Facebook in the privacy of our own boudoirs. Whatever happened to dropping around to see a colleague for a chat, or shooting the breeze sitting in some old armchairs?

Dr Robin Lincoln Wood

Can One Really "Change the World"?

"It had long come to my attention that people of accomplishment rarely sat back and let things happen to them. They went out and happened to things."

Leonardo da Vinci

"Be the change you seek to see in the world"

Mahatma Gandhi

The modern trend of "working on oneself" began in 1859, when the first self-consciously personal-development "self-help" book-entitled "Self-Help", was published. Its opening sentence: "Heaven helps those who help themselves". In classical times, the advice poetry of Hesiod emerged as an early adaptation of Near Eastern wisdom literature. The Stoics offered advice with a psychological flavor. Proverbs from many periods embody traditional moral and practical advice of diverse cultures. A century of insights from the study of our psyche have taught us to explore our interiors as we never have before, unveiling the extraordinary breadth and depth of the human psyche, and highlighting the impersonal, archetypal forces operating in the subterranean chambers of our consciousness.

At the same time, evolutionary biology, cosmology and theology has shown us how to look forward through a clearer set of lenses, revealing the vast evolutionary context in which the daily drama of our lives is unfolding, restoring our faith in the future and reviving a sense of meaning and purpose in our wider world. Evolutionary theory has widespread acceptance and credibility within the scientific world, even if its consequences are still percolating slowly into the popular imagination.

Evolutionary biologist Jared Diamond noted that: "As for the claim that evolution is an unproved theory, that's nonsense. Evolution is a fact, established with the same degree of confidence as our theory of the round earth, our germ theory of disease, and the atomic theory of matter. Yes, there is lively debate about the particular evolutionary mechanisms that caused particular changes, but the existence of evolutionary change is not in doubt"[xxiv]

In this past century we have changed forever our sense of what it means to be alive and human, and our ability to confront the challenges we face in our post-modern society. One could argue that the first Renaissance simply "happened". It is certainly true that no one set out to deliberately create a "Renaissance". But the times were very different 500 years ago, as illustrated by the genre of mirror-of-princes writings, which has a long history in Islamic and Western Renaissance literature. They were either textbooks, directly instructing kings on how to behave, or histories or literary works aimed at creating images of kings for imitation or avoidance.

Often composed at the accession of a new king, when a young and inexperienced ruler was about to come to power, they could be viewed as a species of self-help book. Examples include:

- The Prince (c. 1513)—probably the most famous "mirror". Unfortunately Machiavelli's recommendations regarding the ruthless use of force to maintain power have a great deal in common with Sun Tzu's "The Art of War". Following the recommendations in these works simply leads to increased violence and a ruthless state. Not much fun or virtue there . . .
- Qabus nama (1082)—a Persian example of the genre, which offers considerably more gentle and civilized advice than The Prince—the original is today in a museum in Tehran, Iran. A real pity the Taliban and Al Qaeda never got around to reading this, but perhaps the current President of Iran could be persuaded.
- The III Consideracions Right Necesserye to the Good Governaunce of a Prince (c. 1350)—advice to King John II of France, which it appears he did not read carefully enough, given the disastrous events of his reign. Sadly, being King of England and France at the same time, he ended up beheading too many of the wrong people, and the French nobility no longer trusted him. Cue the "Hundred Years War" between England and France.

As the message on the cover of "Hitchhiker's Guide to the Galaxy" says: "Don't Panic!

The emerging "gen next" (or "generation y") (18-25 yr. olds) together with the even more altruistic "generation x" (26-40 yr. olds) have the potential, if well led, to significantly improve the state of our world. The focus of the Baby Boomer Generation (i.e. most of us between 41 and 60), should be to create the environments in which gen x and y and subsequent generations can create the equivalent of the second Renaissance—all the conditions are there, right now, for this to happen. We should be transcending the narcissism of our generation.

It is easy to become over-focused on our own comforts, our health, growing retirement nest eggs while believing we can save the planet from our armchairs. As a Woody Allen character asserts: "I was born into the Hebrew persuasion, but when I got older I converted to narcissism[xxv]". The Baby Boomer generation should be providing excellent role models, space and resources for those who have the potential to make a real difference "out there".

Moving "away from" climate change, information overload, and high stress lifestyles is not enough to address the core problems of consumerism and the emptiness of the ways of life based on materialism which our electronically accelerated socio-technical system is spreading around the world. We should be "moving toward" a richer, more meaningful existence based on quality connections with each other and our planet, and creating a future our grandchildren can dream about and enjoy. But how?

There are two very different approaches to the world. One is to sit back and appreciate the view as a passenger on life's journey—perhaps a comfortable way to spend a life, but not necessarily the most interesting or satisfying. The second approach is to decide to improve oneself and/or change the world in some way—this becomes more of a challenge, but is potentially very rewarding. I personally enjoy exploring the world, and finding new challenges, yet there is also a part of me that loves being settled, enjoying the comforts of home. Then, just as I have settled in an armchair and am considering such major matters as to whether to replace my worn out old slippers or get another glass of carrot juice, the restless spirit in me calls out again.

There is, one might argue, a third way, which involves thinking and acting as if you are a passenger, but by actually doing what you do you change yourself and the world anyway. Relying on lady luck alone, however, may not have the outcomes one desires.

We are each on our own journey, which though it might bear similarities to the lives of others, is unalterably unique. We can learn from others, and for this we must be thankful, but sometimes there are lessons we need to learn that can only be learned through a medical, financial, relational or spiritual emergency of some kind.

We can set out on a mission to change the world, but at the same time we find that we are, for the most part, changing ourselves. Looking at my own life, I can point to times and places where I have changed parts of the world and other people's lives for the better. Yet there have also been times when the sheer force of events has overwhelmed me, and I've become a passenger swept along by the flow of destiny.

The exit turbulence of the dotcom era at the turn of the 21st century provided me with a front row seat to just such an experience.

Journey through the Eye of a Digital Storm

"In the middle of the journey of our life I came to myself within a dark wood where the straightway was lost".

Dante Alighieri—The Divine Comedy

"Younger adults anticipate that between their late thirties and their early fifties a day will come when they suddenly realize that they have squandered their lives and betrayed their dreams. They will collapse into a poorly defined state that used to be called a nervous breakdown. Escape from this black hole will mean either embracing an un-American philosophy of eschatological resignation or starting over—jaded stockbrokers off to help Mother Teresa, phlegmatic spouses off to the Stairmaster and the singles scene. In short, they will have a midlife crisis."

"Midlife Myths" Winifred Gallagher in The Atlantic, May 1993

We need to rediscover what wholeness (or "holiness") is, in a 21st century context, and what that could mean for each of us without the hollow trappings of crusty religions and rigid belief systems. Perhaps there are really no holy places and no holy people, simply holy moments and moments of wisdom. Finding out who you really are is not easy, and finding out why you are here nigh on impossible in the cacophony of the 21st century. Holy moments are truly exceptional.

Traditional religion has always somehow ended up deifying people, places and rituals. While ritual is useful as a context for personal moments of wisdom, the Hurley Burley nature of modern existence makes it very difficult to align our daily activities and the rituals that might provide us with the space to reflect on our lives, and we end up having to make up our own rituals and find our own holy places in order to experience a holy moment.

Author and Spiritual Teacher Deepak Chopra explains current world turbulence in terms of the newly emerging world centric worldview.

"Whenever there is a faith transition in society, the forces of inertia and resistance also come forth as we rise in consciousness—our shadow also rises to meet the challenge,"

Just as water boils into steam, Chopra says we have the same thing happening in global affairs, "We're seeing the dying carcass of the old paradigm." Yet it may be a little premature to write off the old paradigm just yet—in fact, there might still be a considerable amount of life left in the old boy.

Shortly after the turn of the 21st century and my exit from a "legendary" e-business builder following it being merged with another creature of the same ilk, I headed off for a couple of weeks to reflect on who I was, where I was heading, and what I really wanted to do next. Although the small town of Whistler in the Canadian Rockies is only a two hour drive north of Vancouver, it feels small enough to be on the edge of civilization, smack bang in the middle of the great wilderness of the Rocky Mountains.

Mountains are great places to escape the sterile environment of the modern world. Being close to nature helps us get closer to ourselves. "Little moments of spirit" happen more often, and we open up a bit. I was recovering from my addiction to electronic speed and high technology, through the tranquility of open spaces, and a newfound form of freedom.

Sitting in the airport lounge at London's Heathrow airport, it struck me just how vulnerable we are heading off to a place where you are known by no one, leaving the cosy familiarity of family, friends and one's home town, with all its civilized amenities, far behind. The more

I frequent the hallways and byways of airports, the more I understand their destabilizing power, their ability to say: "My name is Ozymandias, king of kings: Look on my works, ye Mighty, and despair!"

No matter where you are heading or where you have just come from, you are nobody from nowhere until you meet up with somebody who thinks you are someone. This strange alienating quality of the pseudo glamorous shopping malls the other side of departures, grinds with the stark neon reality of the customs halls as you whistle your way through the killing zone where innocent people are stripped of their possessions and their dignity for the sake of a couple of bucks of revenue for Big Brother. Do my travel writing notables like Paul Theroux and Bill Bryson feel like this as they head off on their next adventure?

My own experience of travel is that it levels everyone out. Bumping around inside an aluminum can in the sky certainly puts you all on the same footing, even if the first and business class passengers can feel vulnerable in slightly more salubrious surroundings, with a better quality of empathy from the cabin crew than the sardine class passengers heading for deep vein thrombosis in the middle and rear sections of the aircraft. This was my lucky day—British Airways had given me a free club class ticket to anywhere in the world, and I had to use it up before it expired, so I was headed to Vancouver to meet a few friends and attend a conference.

I am a frequent flyer, and also have a private pilot's license, so it was unusual that the terror of take-off hit me this time in a way I have never previously experienced—just at the last moment, before our taxiing to a stop at the beginning of runway 24, I felt a suffocating feeling creeping up through my solar plexus, engulfing my whole body, until I was ready to rip my seat belt off, run for the emergency exit, fling the door open and jump out onto the tarmac. (I later learn that this is due to uneven cabin pressurization from an empathetic stewardess who says she feels like that regularly). But then, a strange thing happens—an image of my luggage and belongings flying off to Vancouver without me makes me determined to stay on board, no matter how extreme my panic attack.

I now know what Sartre meant in Being and Nothingness even though I could not feel it at the time I read the book as a callow youth—sheer existential terror obviously can creep up on you in middle age, when the confidence of youth has worn off, and the learnings of adulthood are just beginning to percolate through the increasingly unreliable pathways of one's nervous system. As I calmed down and began to enjoy the flight again, I was reminded of the joy of getting one's stomach back after aerobatics flights in my youth when learning to fly light aircraft—a real sense of relief flooded through me and the dry ginger ale tasted like nectar.

From 39000 feet the mountains of Scotland and the glaciers of Iceland looked ruggedly magnificent. Powder puffs of cloud danced over the mountaintops, and the north Atlantic took on a deep, cobalt blue colour. Nature looked decisive and sharply etched, unlike the state of my mind which was decidedly mixed. Small settlements on tiny islands jumped out of the clouds from time to time, pointedly declaring that if you really want to live life on the margins, get yourself a croft on Stornoway, where even small bushes struggle to grow, and trees are smart enough to stay away. What is it about post-modern life, I ask myself? Why, despite the abundance of stuff, and the presence of all modern amenities, is it a clinical fact that despite rising affluence, we are increasingly unhappy?

I may have seen an answer in the snow-capped mountains of Iceland, on which the peaks are slowly disappearing under ever deeper layers of snow, like bizarre pyramids in a desert of ice. We were on the northernmost track I have ever flown, getting as close to the North Pole as I was ever likely to. The bleak edges of the polar ice caps meandered below like some forgotten continent—losing their ice every summer faster than it was replaced during winter. It is now quite possible to believe that global warming will mean we will never see that great flaming yellow ball in the sky over England, or indeed much of Europe, ever again, after the wettest coldest winter since 1679 when weather records began.

Given the roaring background noise of the media, it is now possible to know anything about everything, most of it depressing. Here was a generation growing up with the possibility of what might come close to total knowledge, and with that a potentially commensurate responsibility. Ah, the bliss of ignorance while it lasted!

So, the quest for the Utopia of modernity had many dark sides, which could all come out to haunt you on a bad day or night. This insane struggle for wealth, comfort, release from the daily grind into a fairy tale world of undreamed of riches, could become toxic and dysfunctional very quickly. Yet had I not just consumed a glass of superb Chardonnay and a meal which had managed to rise above the term: "airline food" served by gracious stewards? Was I not now flying effortlessly over the icecap of Greenland, about to enjoy a good nap as a reward for my extreme efforts over the past decade or two? The questions were not getting any easier as I drifted off into a fitful slumber.

The voice of the captain woke me with the news that we had one hour to run before landing in Vancouver. I opened the sunshade to see the jagged teeth of the Rocky Mountains piercing the dense layer of cloud beneath us. Then, as the cloud receded, thousands of tiny pinpricks of light announced the presence of enough water in so many lakes as to make the Sahara bloom. It is obviously true that this northern corner of the North American west coast gets rained on a great deal, even though much of the H2O was still locked up in the snow and ice I would shortly be skiing down.

What makes edges and frontiers interesting is their ability to act as a set of giant question marks. Who are you? Why are you here? Where are you going? And why do you have to come this way? It takes a conscious effort to get to a frontier, and even more effort to figure out what to do when you get there. I take up the challenge to define myself clearly, with no idea of how it is going to work out. But then, life is a path we beat as we walk it.

Canadians and Americans share something very precious: the sense of the last frontier. Here in Whistler you can look out the window of your lodge and stare straight into primeval forests and gargantuan mountains. For North Americans the frontier has taught generations how to go out alone into the wilderness. Going out to be alone raises the ultimate question: who am I? Standing at the peak of a lofty mountain, we are swept with a feeling of awe followed by thanksgiving for being able to be here and to be alive at this moment. Today I can reclaim and repossess myself and take stock of my individual worth.

Yet, despite the wilderness, interactive technologies are hard to escape. They invade our inner peace and occupy our every waking moment. We all need a place to hide, a place to reconnect with ourselves. In many European and Asian societies we find that carefree individuality is a foreign concept. The European or Asian finds privacy in the crowd itself, spending a lifetime learning the strategies and uses of the social mask. The European goes out to be social and comes home to be alone.

The new electronic media have, as McLuhan put it: "begun to shake the distinction between inner and outer space, by blurring the differences between being here or being there." For example, the telephone creates the illusion of being co-present with another person whether they are across the street or on the other side of the world. Now teleconferencing is gradually becoming good enough to do the same thing in a visual medium.

What is identity in a world where I can be known in so many different ways? From personal one-to-one presence, to broadcasting an aspect of one's persona to the world across many media—the conference call, the videoconference, the all-points bulletin e-mail, the television broadcast, the streaming media broadcast across the Internet, and so on. From being a member of a physically co-present team, to being a member of a virtual team; from being a member of a physical community to being a member of a virtual community. In each one of these relationships we participate in very different kinds of conversations with very different kinds of agendas.

Western civilization is built largely upon the extension of one sense, the eye. We have used the eye to create a square framed, controlled environment keyed to the vanishing point on the horizon in the frame. Whether we look at a typical three-dimensional drawing or a two-dimensional screen, we find that Western culture is heavily reliant on a Euclidean space in which representations can be used to abstract the world around us. This has enabled us to create a linear, predictable world where cubes and rectangles dominate in the design of our built environment and where circles and natural forms remain the exception. Anything outside of this artificial order is viewed as nature, wilderness or chaos. So we have created a mechanical world in which we can manipulate objects very precisely, but where we have lost our sense of oneness with the dynamics of the chaotic and complex natural world.

The computer started out as an extension of this linear philosophy, applied as it was to mathematical, statistical and logical problems. As the computer has gradually evolved, it has grown both in power and complexity, enabling it to develop pattern recognizing capabilities. The combination of such pattern recognizing capabilities and networking has created the potential for the global network of computing devices we call the Internet, to become what McLuhan called an acoustic space rather than a visual space.

Where visual space operates in a world of reflected light in which we can discern the existence of multiple discrete objects, acoustic space is a place where resonant and interpenetrating processes are simultaneously related with centers everywhere and boundaries nowhere. We all intuitively understand acoustic space when we listen to our favourite piece of music with our eyes closed—we are suddenly "inside" the music, with the perception that the music surrounds us. If while listening to this music, we hear another sound intruding, both the music and the intrusive sound interpenetrate each other in the same space.

So, if the new media and technology are transforming what it means to be a human being, what can an individual do to ensure their own coherence and quality of life during this transition? One of the great difficulties with the current generation of relatively immature and highly complex technologies is that the uninitiated find it very difficult to improve their own personal situation without significant training and experience.

This is somewhat different to many previous technology revolutions where the individual could master each technology simply by being a part of daily rituals in society. For example, learning how to use the railway was relatively easy for most people in mid-Victorian times—they simply walked to the nearest railroad station and got onto the first train going in their direction. Of course, though it would be helpful to be able to read the timetables, that was not strictly necessary in order to get from a to b.

Going back to earlier times, learning to use language, to hunt and gather, and plough a field were things that were passed down from generation to generation through oral traditions. Formal education began to be required to deal with the more complicated business of learning to read and write and do arithmetic. Now, in the second decade of the 21st century, technology has become so complex and fast moving that even the formal education system is struggling to equip its students with the skills they need to participate fully in the new media. This makes life an interesting challenge for those who have not been equipped with the skills they require to be effective participants in the 21st century economy. Yet, somehow, many are managing to grasp the basics of electronic interaction through a process of trial and error, short courses and a little help from their friends and family.

So, despite fears of a digital divide, the human race is slowly coming to terms with the new media, even if the process is proving to be a little frustrating and quite a challenge for many. We are all gradually becoming part of McLuhan's acoustic space, where everybody can get involved with everyone else in an electric field of simultaneity. As McLuhan himself said: "electronic man wears his brain outside of his skull and his nervous system on top of his skin . . . he is like an exposed spider squatting in a thrumming web, resonating with all other webs. But he is not flesh and blood; he is an item in a databank, ephemeral, easily forgotten, and resentful of that fact. Earth in the next century will have its collective consciousness lifted off the planet's surface into a dense electronic symphony where all nations—if they still exist as separate entities—may live in a clutch of spontaneous synesthesia, painfully aware of the triumphs and wounds of one another. After such knowledge, what forgiveness"[xxvi].

As we race towards this instantaneous apprehension of the totality, we run the risk of losing touch with our own experience of nature and of ourselves. Half of the planet has a mobile phone, 40% use a computer and 30 per cent use the Internet. We are connecting humanity so that we can communicate from anywhere with anyone as well as conduct much of our business, personal affairs, learning and entertainment on the move. The transformation in the way we live our lives and the danger is that we will not only find it hard to switch off—we might find that we are losing touch with our common sense, our unique human power to translate one kind of experience in one sense into all other senses and interpret that as a coherent experience.

Against this backdrop of global omnipresence I find myself staring at a two-dimensional screen talking into a microphone converting my spoken words into reams of text, and looking out the window into virgin pine forests. I have just done my e-mail, checked my voice mail. I've spent 30 years tinkering with computers and believing that they really will change the world. But a part of me also secretly sees the world from another perspective. Many people would much rather have "reality reality" than "virtual reality". Although the digerati may assert that being digital is just the thing, being physical is a much better thing. Many of the people in computing, telecommunications and the new media genuinely believe that we should have our eyeballs glued to a two-dimensional screen simply because that is how they make their money.

Mountain peaks are big contenders for holy places. Mountain peaks covered in deep snow are even better. The pure whiteness of a mantle of snow swaddling hundreds of jagged peaks is in itself a religious experience. On a sunny day when that big yellow ball of fire beams down from the blue vault of heaven onto pristine powder snow, it is difficult to imagine a better place for a holy moment. Skiing through the dappled beams of light dancing on deep powder snow between fir trees exuding their pungent menthol aroma, is in itself a form of deep prayer and meditation.

Of course we cannot remain on the mountaintop forever. We must eventually return to the streets and the maddening crowd. Even here though, it is possible to find a kind of peace. Just a few steps from the Place de la Concorde in Paris lies the Buddha Bar. Founded by a dreamy Parisian Raymond Visan, the Buddha Bar is designed to be a haven of peace in the noisy, chaotic heart of one of the world's great Metropolises. The high ceiling evokes an Asian temple. An immense Buddha towers over the room. The DJ is playing mysterious music expressing the delight and magic of Indian, Chinese, Arabic and other melodies. The parties, meetings and exchanges taking place are full of energy. The Buddha Bar is a place which makes one feel good, as it takes its clients on a voyage into the world situated between dreams and reality.

Even in the more confined spaces of a Starbucks or a coffee bar of that ilk, a frothing cappuccino with chocolate sprinkled on top can come close to an epiphany. The incredible popularity of places which serve expensive caffeinated beverages appears to lie in the opportunity they provide the weary traveler for a few moments of respite from the hectic activities of the day. Sipping a latte with the steam coming off it, it is possible to understand what Lie Tzeu said about dreams: "The ultimate dream of the traveler is to stop knowing what he is contemplating. Every being, everything is an occasion for travel, for contemplation".

Of course it's not always that easy to get a high. Or more importantly after a high you can almost guarantee some kind of low. Three days of hard skiing left me feeling as floppy as a large slab of jelly. I woke feeling so out of sorts I spilt enough grapefruit juice on my computer keyboard to completely disable most of the keys. It is difficult to think philosophically when your computer does not understand you—voice recognition software is still a long way from being perfect.

Ask Stephen Hawking, the eminent English cosmologist who is afflicted with a severe degenerative disease of the nervous system which means he has to translate movement in one of his hands into speech through a computer program. Thanks to a new computer program, after years of speaking in the halting American accent of the original speech synthesizer, he can finally express himself in good Cambridge English. Without the keyboard I can just barely begin to understand the level of frustration Hawking must experience whenever he wishes to express himself. Though my present affliction is, in comparison, trivial in the extreme, it is nonetheless a humbling experience to have to repeat almost every third word. It makes one realise just how miraculous human speech and hearing really is, and just how clunky screens, mice and keyboards are.

Yet, in the midst of and enabled by this techno-trivia, I look out the window at *reality* reality. This has to be one of the most beautiful and inspiring places I have ever been to. There is still the smell and feel of the wilderness here, and the friendliness of the frontier spirit. John Katz's book: "Journey to the Mountain" was one of the inspirations for this trip. He writes about his own journey of faith and change, and how he found out who he was and renewed himself when he turned 50. I feel like this is a watershed for me for me too. Little did I realise how profoundly my life would change in less than 12 months. I was being given the opportunity to realise my true destiny, but first I needed to cleanse my lenses of perception to see more clearly into my own heart and soul.

Getting out of the rat race for a couple of weeks to reconnect with nature and oneself may in itself only be a small gesture in the grand scheme of things, but the world would be a very different place if we could all find that still and sacred place inside ourselves from which we discern our destiny. I was like a small boy again—not being able to wait for the sun to come up and get out there to find out what happens next in this glorious thing called life.

It was my last day in Vancouver. I had left Whistler on a bus almost one week before, and met up with a friend, who invited me to stay the night at his home in a quaint seaside resort town called White Rock. It reminded me of the better south coast British seaside resorts, without their seamier sides. His home was a haven of tranquility, and his family charming. I felt possibility space all around me in the open skies, the light spacious rooms, and the easy-going atmosphere. We discussed complexity, philosophy, people and nature. We played songs from the sixties and seventies on two guitars, and sing in our out of practice "guys who used to have bands when they were young" voices. I explored the town of Victoria on Vancouver Island, a delightful admixture of British colonial and native Indian culture and architecture. The tall totem poles were particularly fascinating, as they speak to something deeper inside me about life and the universe.

My friend had also spent the past decade re-examining his work, life and contribution to the world. He discovered that while he used to design physical hospitals, he was now solving the more complex problems of the healthcare system, and the social design of hospitals and healthcare, while developing an understanding of what makes communities tick, and how effective teams deliver great results in a healthy organizational environment. And to do this he has also undergone a major personal journey that, in some ways, mirrored my own. Perhaps there is something to learn here about creating sustainable human systems that are a pleasure to live in?

From all of this, at a basic level, I had understood that paradise was something like a fabulous table in a decent restaurant with a sea view, accompanied by a bird orchestra in the palm trees, in a warm, friendly part of the world, not too far from mountains where skiing in powder snow was readily available. Of course a good book, fine music, culture, friends, and the love of one's life, would complete the picture.

But where was such a place to be found in the maddening hurly burly of the 21st century, especially when there was still so much suffering to go around?

Dr Robin Lincoln Wood

The Law of Threes

"O human race born to fly upward, wherefore
at a little wind dost thou fall"

Dante Alighieri—The Divine Comedy

I believe these are exciting times to be alive, and that we are all part of a great historical shift which is already underway. For the first time in history we are evolving a genuinely global perspective amongst the leaders and intelligentsia of the more developed nations on our planet, which provides us with a unique opportunity to mobilize an enlightened network of innovative leaders who can shake things up for beneficial change.

That is what I believe on a good day. However, the trouble with paradise is that not every day is as good as it gets.

Hindsight really is 2020 vision. Major shifts and changes in life rarely announce themselves beforehand. Shortly after the turn of the 21st century, having feasted at the fin de siècle buffet of the digital Belle Époque, all hell broke loose. For anyone living in the modern world near a television screen on September the 11th, 2001, the collapse of the Twin Towers was the first sign that nothing would ever be the same again. In fact, the law of things happening in threes appears to have applied itself with a vengeance to many lives within months of this tragedy.

My own life was no exception to this rule of threes. In the months leading up to 9/11, several of the startups I had invested in and advised during the birth of the digital economy, failed. The internet company of which I was a managing director was taken over, and I was given a severance package. I was lucky to find a few part-time roles, as a non-executive director and teaching top executives strategy at the London Business School. There were board meetings, presentations and conferences in places like Paris, New York, London, Boston, San Francisco, Monaco, Brussels and Prague. But times were tough, and everything I had invested both financially and energetically into a future nest-egg for myself and my family, disappeared overnight.

Then, one week after 9/11, after years of heading in different directions, my wife and I agreed on an amicable separation while sitting in our local Starbucks. While my head knew this was the right thing to do, my heartache at the failure of our marriage and the implications for our two children knew no bounds. Breaking up always hurts, but separating a family is one of the toughest decisions to make. A second blow had struck.

Just two weeks later, in Sandton, the business center of Southern Africa north of Johannesburg, the third blow struck without warning: my father dropped dead of a sudden heart attack a few months shy of what would have been his 68th birthday. Ray was getting ready for his ritual Sunday morning game of tennis at the Bryanston Country Club, where he had played for 25 years. He had been a mentor and role model for myself, my brother, sister and many others of his own and our generation. He is remembered as a visionary leader, publisher and businessman who had dedicated his life to building a new and better South Africa through both business and philanthropy. And now, he was gone, long before he could put his feet up and enjoy his grandchildren.

Yet in the midst of tragedy, life continues. After joining family and friends for my father's funeral, I returned to London from Johannesburg, shaken, but determined to continue my own life's work. My new book "Managing Complexity" was receiving awards, and I was invited to address the policy advisors at number 10 Downing Street on the implications of the digital economy for social capital. The thrill of presenting to Tony Blair's leading advisors brought a welcome ray of sunshine into an otherwise dark and stormy picture, and the start of an exciting new program with the world's leading mobile telecommunications company at London Business School helped me forget, briefly, the deep pain I felt burning silently in my gut.

During the 1990's a decade of global growth created the hope that most of the world's big problems could be solved through better management and fine-tuning of the capitalist system. The wave of privatizations unleashed around the world first in the UK, then New Zealand and Australia, Eastern Europe, Russia, South America and Southern Africa since the mid 1980's had indeed created more efficient, value-adding organizations and services in everything from Post Offices to railways to water companies. Despite many excesses, much greed amongst those grabbing assets at a discount, and spectacular failures by British Rail to deliver a safe, reliable service, the free-market appeared to have transformed a great deal of the economy and given consumers a better deal. That was certainly the case with British Airways which went from being one of the world's worst airlines in the early 1980's to one of the best in the 1990's.

The dotcom era and the first stirrings of the digital economy in the late 1990's held an almost messianic allure for many. Technology, innovation and a new generation of entrepreneurs were remaking the world, and the dinosaurs of the bricks and mortar economy were now giving way to the rock-stars of the clicks and mortar economy. Sitting in the old factory of Andy Warhol in Union Square in New York City amongst my e-business colleagues, there was a sense of destiny that required us to look beyond how things were done today to the radical shifts that would become possible when the world became fully wired and connected.

The digital revolution was and is truly global, and has transformed parts of China, India, the tiger economies of Taiwan, Singapore and Malaysia and even southern Africa, in addition to North America and Europe. The one laptop per child project of the MIT Media Lab promised that everyone would have access to this powerful new technology, everywhere, and this may even come to pass eventually.

There was an undeniable euphoria in the air as the Berlin wall came down in 1989, and Francis Fukuyama boldly declared the "End of History"; meaning that the free market system had won against the ideologies of communism and socialism. As I write this I am looking at a small section of the Berlin wall encased in Perspex, which brings back those halcyon days when it appeared that democratization and capitalism would finally bring about a better world for all. Indeed, the Rio Earth Summit in 1992 promised many great things, and delivered on quite a few of them: the ozone layer, rainforests and the Kyoto Treaty. Sadly, it could not deal with the excesses of consumerism run amok and governments reluctant to lose votes by demanding that their citizens begin to respect the environment, save energy and reduce their levels of conspicuous consumption.

My own experience of the era that began with the election of Margaret Thatcher as Prime Minister of the UK in 1980, and ended with the fall of the Twin Towers on 9/11/01, was that the rising tide of science, technology, business, education and economic development that characterized the latter half of the twentieth century was just not enough to transform our world. During those two decades the promise that the rising tide of ever increasing economic growth and globalization would lift all boats equally, had proven to be spectacularly false.

Indeed, the rich had got richer, the ranks of the middle classes had grown worldwide, but the bulk of poor people around the world had generally been left behind. If the ancient challenges of humankind were not to be resolved by such a massive wave of global development driven by increasingly turbulent markets prone to spectacular bubbles and crashes, then what were we missing? Although poverty, war, disease, dictatorships, droughts and famines have always been with us, there was a glimmer of hope that we were going to get on top of these scourges through the Millennium Development goals, hundreds of thousands of NGOs and increasingly enlightened businesses and governments.

But the twin rhetorical devices of free markets and democratization that propped up right wing and center right governments worldwide failed dramatically in their own ways between 1 January 2000 and the end of the first decade of the twenty-first century. The collapse of the two largest economic bubbles in history, the failure of two of the largest foreign invasions and wars ever conducted in Afghanistan and Iraq, along with the fall of two of the tallest towers ever built in New York, did not bode well for the future of our species. At the same time, the grim specters of climate change and mass species extinctions confronted us at exactly the same moment we entered the twin peaks era of peak oil and peak soil.

I was definitely in need of a break from all this. In conversation with a friend, she suggested that it might be time to dust off an old dream I had always harbored, of creating a little corner of paradise in a sunny, peaceful part of the world between the mountains and the sea.

L'Art de Vivre

"A great flame follows a little spark."

Dante—The Divine Comedy c. 1300

The ancient, highly polished Citroen 2 CV ("deux chevaux" or "two-horses") wound down the mountain lane toward the Spanish border as the sun set over the Pyrenees. I sat in the passenger seat beneath an open canvas sun-roof as my friend down-shifted around a tight corner. We were on our way to a magnificent Catalan dinner outside La Jonquera, just the other side of the French border in Spain, to celebrate my purchase of a derelict but charming Belle-Époque chateau in Perpignan.

"Laurent", I ventured, "I see you like old cars", as we drove slowly down the highway, being overtaken by every other car on the road. "Yes", he replied, "They remind me of happy times and help me to slow down". Having spent 30 years in the fast lane myself, I had experienced the downside of speed. I had won and lost several times. Yet through all these ups and downs I had simply assumed that stepping off the merry-go-round was not a real option. Slowing down was something to do in my retirement, not my mid 40's.

"L 'art de vivre", as the French say: "The Art of Living". While something of a bon viveur and connoisseur of the finer things in life, I had always wanted to take some time out from the rat race in the Big City. Reading Zen and the Art of Motorcycle Maintenance twice, meditating regularly, learning about personal transformation, working on many organisational transformations and attending several personal development programs had been a good start.

But only now was the real message sinking in: I could spend every evening for the rest of my life watching one spectacular sunset after another, and still never get bored with this place. L 'art de vivre—that was definitely something I could use a lot more of . . . .

I had landed in Perpignan, the ancient capital of the Kingdom of Majorca, a town of 150 000 sun-drenched inhabitants perched between the mountains and the sea, 30 km from the Spanish border in southwest France. The ancient streets are clean, one can walk around perfectly safely anytime of the day or night, and people acknowledge each other in the streets and shops—for example, when going into the local boulangerie or the doctor's surgery, one says "Bonjour", and everyone in the queue or waiting room echoes "Bonjour!" back. Not a major thing, on its own, but when combined with other small courtesies, the cumulative effect begins to add up to a society which feels warmer and friendlier than those in chillier northern climes.

I was curious—where did this kind of culture, so different to the high-speed, clinical and calculating cultures of northern of Europe and the Americas, come from? Of course, we were in "Le Sud", the south, where the sun shone, people ate outdoors and mingled in a leisurely fashion on the streets. This Mediterranean dimension definitely played an important role. But there was also a unique flavor to this Mediterranean space that seemed much more ancient than contemporary French or Spanish culture. It could be seen in the architecture, the sound of the local Catalan tongue, itself descended directly from the Romans.

Here, in warm southern Europe along the Mediterranean coast, a very different version of history had played out over the past millennium. The Catalans had established their identity over one thousand years ago, starting with the Kingdom of Mallorca which was founded in the 11th century, with Perpignan as its capital and James I as its king.

A wave of Moorish art, music, medicine and culture moved up through Spain from North Africa, along with the re-conquest of Spain between 790 and 1300. Although the Moors were finally defeated in 1492 at the Alhambra, their architecture and design influenced subsequent generations until the present day between Granada and Perpignan. With the exception of the magnificent Moor, Othello, in Shakespeare's play of that name, things Moorish were to disappear from the radar screen of western civilization for a very long time indeed.

When James I died in 1276, the Crown of Aragon passed to his eldest son Peter, known as Peter III of Aragon or Peter the Great. The Kingdom of Mallorca passed to another James, who reigned under the name of James II of Mallorca. The kingdom included the Balearic Islands: Mallorca, Minorca, Ibiza and Fomentera. Today we associate Ibiza with all night raves and clubbing, Mallorca with the more mature jet set such as Michael Douglas and Catherine Zeta-Jones and other celebrities who own properties on the island, while Minorca and Fomentera tend to stay out of the limelight.

The king was also lord of the counties of Roussillon and Cerdanya and the territories James I kept in Occitània (now Montpellier). The legacy of James I was significant, resulting in the unification of previously warring territories.

In 1279 the Treaty of Perpinyà created an imbalance of power between the Kingdom of Aragon and the Kingdom of Mallorca was created. James III of Mallorca took the throne at the age of nine. In 1325, the regency council secured the renunciation by the Aragonian king of any claim on the rights of succession of the Mallorcan throne after the repayment of a great debt incurred by Sancho I during an invasion by Sardinia. While this act solved the problem of succession, it also plunged the kingdom into a serious financial crisis. In May 1343, Peter IV invaded the Balearic Islands, and followed that up with the invasions of the Roussillon and Cerdanya.

James III was only able to keep his French possessions. After the sale of these possessions to the king of France in 1349, he left in a huff for Mallorca. With this, the Kingdom of Mallorca was definitively incorporated into the Crown of Aragon. To the north, the Pope and various French kings held sway from Paris, Rome and Avignon. To the south, the Moors, then the Aragons and Castilians ruled. By 1500 the French had taken control of northern Catalonia, and the Spanish ruled southern Catalonia to Barcelona, Lleida and beyond.

Europe had now reached a turning point after the feuding and wars of the dark ages. The start of the 500 year "age of reason" in 1500 unleashed massive creative and productive forces which changed everything: art, religion, politics, science, technology, music, manufacturing and commerce. The fruits of the first Renaissance were many and varied—from our democratic systems of governance, to unprecedented material progress, to laissez-faire culture, benefitting unimagined numbers of people. The art, music, science and light technology of the Renaissance spread all the way from Florence to Gibraltar, with Perpignan lying midway between the two along the ancient Roman road, the Via Domitia.

A Taste of Paradise

> "If you follow your bliss, you put yourself in a kind of track that has been there all the while, waiting for you; and the life you ought to be living is the one you are living. Wherever you are—if you are following your bliss, you are enjoying that refreshment, that life within you, all the time."
>
> *Joseph Campbell*

The origins of the word Paradise lie in the ancient Median language of Persia, where a paradise was a walled in compound or garden. In

English the name is now commonly used as an alternative for "heaven", while New Agers, personal development fanatics, mystics and some old hippies refer to this state as being "blissed out".

The Persian paradise garden is one of a handful of fundamental original garden types from which all the world's gardens derive. In its simplest form, the Persian garden consists of a formal rectangle of water, with enough of a flow to give it life and movement, and with a raised platform to view it from. A pavilion provides shelter and formally arranged trees provide shade, while the perimeter is walled for privacy and security. Odour and fruit are important elements in this "pairedeza" or paradise, which realizes the symbol of eternal life, a tree with a spring issuing at its roots.

The grounds of the Taj Mahal, the tiled courtyards, arcades, pools and fountains of Moorish Andalucia, the designs for the Versailles and Louvre Gardens—all are based on the original paradise gardens of those ancient Persians. The original site of what we now call the Garden of Eden is said to be located somewhere between Iraq and Iran, with the Euphrates and Tigris rivers featuring prominently in Biblical accounts.

Human nature being what it is, hundreds of versions of this mythical garden and its location abound, the most colourful of which must be that held by the members of The Church of Jesus Christ of Latter-day Saints (also known as the Mormons). For the Mormons, the Garden of Eden is believed to have been located in present-day Jackson County, Missouri, on American soil. According to their theology, Independence, Missouri, was revealed to be the "centre place" of Zion and the original dwelling place of Adam and Eve in the Garden which God planted "eastward in Eden".

It appears that there is a deep need in the human psyche for a perfect, paradisiacal place where there is an abundant supply of all our needs and a safe place to rest our heads when our work is done. Indeed, psychologists have had a field day with this particular idea, particularly the German-American Jewish social psychologist, psychoanalyst and humanistic philosopher Erich Fromm. Here is what Wikipedia has to say about this:

"The cornerstone of Fromm's humanistic philosophy is his interpretation of the biblical story of Adam and Eve's exile from the Garden of Eden. Drawing on his knowledge of the Talmud, Fromm pointed out that being able to distinguish between good and evil is generally considered to be a virtue, and that biblical scholars generally consider Adam and Eve to have sinned by disobeying God and eating from the Tree of Knowledge. However, departing from traditional religious orthodoxy, Fromm extolled the virtues of humans taking independent action and using reason to establish moral values rather than adhering to authoritarian moral values.

Beyond a simple condemnation of authoritarian value systems, Fromm used the story of Adam and Eve as an allegorical explanation for human biological evolution and existential angst, asserting that when Adam and Eve ate from the Tree of Knowledge, they became aware of themselves as being separate from nature while still being part of it. This is why they felt "naked" and "ashamed": they had evolved into human beings, conscious of themselves, their own mortality, and their powerlessness before the forces of nature and society, and no longer united with the universe as they were in their instinctive, pre-human existence as animals. According to Fromm, the awareness of a disunited human existence is a source of guilt and shame, and the solution to this existential dichotomy is found in the development of one's uniquely human powers of love and reason."

Luckily for the Catalans, they were already living in the Garden of Eden, and unlike the rest of human kind, were not desperate to get "back to the garden", as those words were immortalized in an iconic Crosby, Stills and Nash song of the late 1960's. Not only that, but the entire ugly, dehumanizing apparatus of the Industrial Revolution which so disfigured other countries and regions, simply passed Catalonia by. Today one can still find a very strong relationship between nature and the Catalan people, embodied in everyday activities from small-scale farming to growing wine, peaches and various other fruits on every inch of fertile soil from the shores of the Mediterranean to the tops of the Pyrenean foothills.

On a typical summer's day where I live, the lizards are out under the climbing jasmine, going about their business of bug hunting. I feel welcomed by their ease at the Chateau, scampering across the terrace in the morning sun. At night, the chameleons prey on fluttering moths gathering around the courtyard light at dusk, deceptively quick as they close in on their prey. After dusk the lights of the exotic gardens switch on, illuminating the Zen cascades, white stone paths and marble walls.

We decide to take an evening swim in the azure pool, while two small bats whirl around each other in the twilight.

Here in the shadow of the mountains and amongst the ancient vineyards of French Catalonia, life flows in a dance of heat and light. All around us, the grapes swell in ripe bunches that will soon ferment into a golden wine filled with summer warmth. In 1750 Thomas Jefferson ordered 150 cases of the sweet dessert wine "Muscat de Rivesaltes", which grows a stone's throw from where I am sitting. Although Jefferson clearly had a sweet tooth, there are hundreds of varieties of Languedoc-Roussillon wines, in a region which produces two-thirds of the wine of France. We will return to this important topic a little later in this book.

Cicadas pierce the silence with their chatter and wild lavender and mint yields up its scent. Above all, the sun presides over the mountains rising to dizzy heights across the plains: a hazy wedge between land and sky. From this distance the only hint of bustling resorts in the grip of August are the cars that ply the highways to the coast. Here, we share our palm trees with the lizards.

The rhythms of our day slow to a less frenetic beat and the kinks in our neck, shoulders and back muscles are soothed by the sun's deep heat. After the confines of the city, the sense of space and freedom is immense. We alternate days amongst the tourists on the local beaches with secluded swimming holes high in the gorge behind Sorède. We enjoy lazy beach picnics and then wallow in the sea or splash through freshwater pools, shrieking at small fish that tickle our toes.

When the *Tramontane* is howling, one can head for the foothills and the ancient village of Castelnou or the Museum of Modern Art in Ceret where we marvel at the brush strokes of Matisse and Picasso ceramics so delicate in form yet so violent in bull fighting imagery. The valley fills with mist and we drive across to Spain to find sun and surrealism. It is fabulous fun to flit across borders and in the Teatro Museo Dali in Figueres one encounters a magnificent and inspiring monument to the greatest of the Surrealists, as if one was in a day long daydream.

The single track of Le Petit Train Jaune clings precariously to the mountainside as it pulls us high into the Pyrenees from Villefranche de Conflent to Font Romeu. We snake our way up the passes in a steady climb over bridges, streams, villages and roads to the edge of the tree line. It feels cold when we stretch our legs round the ski resort on top—until we realise this is comparable to an English July. We rattle back down as an autumnal dusk settles with the smell of wood smoke in these elevated valleys and resume our summer in Perpignan.

We are spoilt by the rich selection of local produce in the markets and shops. Seafood and charcuterie, ripe cheese, heavenly bread, snails and remarkable wine make this a quintessentially Catalan affair. A ten minute walk away in medieval Perpignan, amongst the hundreds of restaurants, we enjoy a range of gastronomic delights. It is hard work living in Paradise, but someone has to do it.

I am a sun worshipper. Thanks to a fantastic local microclimate, (and, regrettably, global warming), of a winter's lunchtime I can take my shirt off and spread my back out against the firm, dark hardwood of my Adirondack armchair on our sun porch, which even in the middle of the European winter can reach 25 o C+. I look out over a view that would make the designer of the Garden of Eden envious: several different species of palm trees, pink, red and white camellias in blossom in January, lush expanses of green lawn, and the carefully tended remnants of a classic French chateau garden. But one can be just as happy in a small backyard in London or Timbuktu, providing the sun and nature are there too.

This particular garden represents a labor of love of many years duration. I inherited a jungle from the previous owner, who was 95 and had done nothing since 1945 except pull out a few weeds. Close to 60 years of nature run wild had led to the "bad" species running amok over the "good", together with the natural ageing of trees and shrubs. The chainsaws came out together with serious German garden trimmers, and two years later the jungle had been tamed. The next step was to design our own version of paradise. Strongly influenced by the Moorish approach to fountains and running water in courtyards, along with the naturally good feng shui of Chinese design and some lessons from Zen gardens, we set about incorporating all the beauty we knew of into a place that would feel like Eden, but better, and without the snake.

The soil is incredibly fertile—the flood plain of the River Tet which runs down from the snows of 2 784m Mont Canigou visible from our tower, benefits from some of the best top soil in France. Everything grows with gay abandon, organically, without a hint of man-made substance in sight. It is like owning a nursery, with lilies of the valley, palm trees, bamboo and exotic plants seeding themselves everywhere. There is a lot of weeding, pruning and cutting back to do, but the sight and scent of a dozen red roses dangling over the marble fountain together with the sound of the Zen cascade burbling in the background, make it all worthwhile.

The birds have given our acre of paradise their seal of approval too—from turtle doves to seagulls to sparrows, finches, swifts and robins, there is an endless parade of avian magnificence flitting from branch to branch, and getting their beaks wet in the numerous water features and fountains. The garden is protected by ancient stone walls of over 6m (20 feet) in height, and is crowned at the far end by an azure pool and 2m x 2m (6 foot) waterfall. The water comes burbling out of the ground from a borehole 27m below the surface, and is of mineral water quality. Mother Nature looks after us very well here, and we do everything we can to return the compliment.

Here, then, is a simple recipe for bliss: take one human body, add a twist of sunshine, and then garnish with a dash of nature. For extra special happiness, add a loving mate and a drink of your choice. Splashing in a bit of water is also highly recommended, and if you can find a pair of turtle doves to coo you are in paradise. The smells of jasmine, lavender and orange blossom complete the picture.

But just like Eden, it does all seem a little too perfect. Might a worm be eating its way through the heart of this delicious apple? Might the Barbarians storm the gates as the rising oceans drive the starving masses to desperate acts? If the combined might of science, technology, business and the worlds peace, humanitarian and aid movements were not able to stitch up the wounds of planet earth, what could? Perhaps we had we been focusing too much attention on the outer state of humankind, and not enough on the inner frontiers that psychology had begun to open up more than a century ago? Beyond our opinions, attitudes and beliefs as measured incessantly by polling organizations, might there be a deeper layer of motivational challenges and barriers to change?

Can A Theory of Everything Change Anything?

In October of 2001, I found myself on a flight to Denver, Colorado to attend a week long course in the Rocky Mountains with a tall, genial Texan named Dr Don Beck. In addition to his lifelong immersion in the field and practical experience as a Professor of Psychology, Dr Beck's claim to fame is that he successfully applied social psychology to help bridge the massive gap between right wing Afrikaners and Black Power ANC supporters in the peace process in South Africa between 1984 and the election of Nelson Mandela as President in 1994.

I discovered Dr Beck through a friend and colleague I had spent several years working with in Asia. My friend had been engaged to help one of the world's oldest corporations prepare for the handover of Hong Kong to China in 1997, and Dr Beck's book, "Spiral Dynamics" was published in 1996, just one year into our collaboration. I brought to this endeavour every management and leadership tool that had proven effective in my own work since 1975, together with their integration into the diagnostic framework called "The Change Wheel"[2].

Dr Beck had also teamed up with a controversial philosopher whose claim to fame is his "Theory of Everything". Tall, with a shaven head, gaunt yet muscular physique, and a wry sense of humor, Ken Wilber's dozens of books on life, the universe and everything are still among the best-selling popular philosophy books in the world. Wilber's claim to fame is that he has successfully created a philosophical "operating system" which integrates all quadrants of existence with all levels of existence with every line of human development and every state of human experience. (What is known as the "All-Quadrants, All Levels" framework, abbreviated as "AQAL").

Don Beck's life's work had been strongly influenced by two legendary psychologists: Dr Clare Graves and Dr Muzafer Sherif. An extraordinary social psychologist, Dr Sherif was a founder of modern social psychology. Sherif developed several unique and powerful techniques for understanding social processes, particularly social norms and social conflict.

Sherif is famous for the Robbers Cave Experiments. This series of experiments, begun in Connecticut and concluded in Oklahoma, took boys from intact middle-class families, who were carefully screened to be psychologically normal, delivered them to a summer camp setting

---

[2] For which I was awarded a doctorate in 1995 by London Business School following a decade of action research in the world's leading corporations, banks and governments.

(with researchers doubling as counsellors) and created social groups that came into conflict with each other.

These studies had three phases. The first was "Group formation", in which the members of groups got to know each other, social norms developed, leadership and structure emerged, and then the second stage began: "Group conflict", in which the now-formed groups came into contact with each other, competing in games and challenges, and competing for control of territory.

The third and final stage was "Conflict resolution", where Sherif and colleagues tried various means of reducing the animosity and low-level violence between the groups. It is in the Robbers Cave experiments that Sherif showed that superordinate goals (goals so large that it requires more than one group to achieve the goal) reduced conflict significantly more effectively than other strategies, such as communication and contact.

In contrast to the avuncular Turkish nature of Dr Sherif, Dr Clare Graves was a very deliberate, gentle giant whose life's work had been carried out at Union College in Schenectady, New York. There he developed an "epistemological model" of human psychology. Graves claimed that the inspiration for so doing came from undergraduate students in his introductory psychology course. He acknowledged that he was unable to answer the frequently asked question as to who from among the many competing psychology theorists was ultimately "right" or "correct" with their model since there were elements of truth and error in all of them.

A number of theorists have been influenced by Graves' "Emergent Cyclic Levels of Existence Theory". Chris Cowan and Don Beck used it as the basis for their book Spiral Dynamics: Mastering Values, Leadership, and Change, which in turn is referenced by our integral theorist, Ken Wilber in his later works. Dudley Lynch has used it as the basis for four books, including The Strategy of the Dolphin: Scoring a Win in a Chaotic World (with Paul L. Kordis). John Marshall Roberts builds on Graves' work in his book Igniting Inspiration: A Persuasion Manual for Visionaries. Many other writers and consultants find merit in the Gravesian perspective in domains ranging from personal coaching to executive assessment, from organization design to social policy.[xxvii]

So herein lies the challenge: if the Trouble with Paradise is partly caused by clashes between people with different value systems, and partly by the inherent tendency of human beings to polarize issues and situations, then what (as another social philosopher, Karl Marx, asked a few centuries ago), is to be done?

It would take me over a decade of research and development in my attempts to apply these insights in my own life and work, to arrive at some tentative conclusions to this hard question. What follows is a brief account of what I have learned on my journey.

Chapter Four
The Trajectory of our Species

"There is grandeur in this view of life, with its several powers, having been originally breathed into a few forms or into one; and that, whilst this planet has gone cycling on according to the fixed law of gravity, from so simple a beginning endless forms most beautiful and wonderful have been, and are being, evolved."

Charles Darwin—On the Origin of the Species by Means of Natural Selection of the Preservation of Favoured Races in the Struggle for Life

"We are evolution incarnate. We are her self-awareness"

—Howard Bloom

In the 21st century, to answer the question "What is to be done?" (one also posed by Karl Marx in response to the social horrors of the Industrial Revolution), we need to rewind the clock back to a time when there were not only no clocks, but no such concept as time itself. We need to ask not only what it means to be human, but also how humankind is evolving and could evolve.

Today, over half of the human species is living in the cities and urban areas of the world. These urbanites have managed to distance themselves from nature, and the world in which our ancestors evolved. Because evolution is now being driven by cultural and technological forces more than by biological forces, it is easy to imagine that we have escaped all the pain, suffering, and difficulty inflicted by being just another species rather than the top predator in every conceivable food chain on the planet.

We also face an eminently practical problem. Due to our genius as a species, we have hit the limits to growth of our industrial civilization on our planet. There are few who have not heard that we face an ecological crisis, an economic crisis and a social crisis, all at the same time. For human beings, a temperature increase of 6 degrees Fahrenheit is the difference between life and death—for our planet a temperature increase of 6 degrees centigrade is the equivalent of a death sentence for 90% of species, including us.

Since the turn of the 21st century, we have been living on borrowed time, using the resources of 1.4 planets to fuel the rapid growth in the developing economies and the greying, retiring populations in the developed economies.

It is essential that we deal with the problems of evil and suffering, which have preoccupied so many great philosophers, wise men and women, and spiritual leaders over the ages. Essentially, we all have a deep need to understand why bad things happen to good people, why only the good die young, why life appears to be a veil of tears and suffering for so many, and so on. This is no mere academic matter, but a central issue for all sentient beings who are able to reflect on their own predicament and make choices.

To understand the emerging story of what it means to be alive in what is still a very mysterious universe, living in very large numbers on a highly paradoxical planet, at a critical moment in the evolution of our species, we need to dig a little deeper into what drives evolution.

For 14 billion years or so literally everything that is required to enable life to exist on planet earth, from the crust to the oceans to the atmosphere to the weather to the biosphere—(in short, "Gaia"), has been evolving unconsciously. This led to the appearance of the first single-celled organisms 4 or so billion years ago, then to plants, animals and mankind.

In order to appreciate the many critical moments in the evolution of life, who better to ask than a mitochondrion[xxviii], one of the earliest forms of life still going? There are thousands of these tiny organelles in every single one of the hundred trillion cells in your own body, so finding a talkative mitochondrion should not be too difficult.[xxix] Not only that, mitochondria demonstrated the most important principle in life and the evolution of life: that collaboration is a much more intelligent strategy than competition, especially during crises. We will also see in this short story that crises themselves are the most fundamental drivers of evolution.

In the Beginning . . . was the Mighty Mitochondrion

Instead of a slow, continuous movement, evolution tends to be characterized by long periods of virtual standstill ("equilibrium"), "punctuated" by episodes of very fast development of new forms.

Stephen Jay Gould

Scientists estimate that life on earth began around 4 billion years ago. In the energetic chemistry of early Earth, a molecule gained the ability to make copies of itself, becoming a replicator. In making copies of itself, the replicator did not always perform accurately: some copies contained an "error." If the change destroyed the copying ability of the molecule, there could be no more copies, and the line would die out. On the other hand, a few rare changes might make the molecule replicate faster or better: those "strains" would become more numerous and "successful." As choice raw materials[xxx] became depleted, strains which could exploit different materials, or perhaps halt the progress of other strains and steal their resources, became more numerous.

These were the early ancestors of the mighty mitochondrion, which still animates all living creatures today. If the hundreds of trillions of mitochondria in your body stopped working right now, you would rapidly grind to a halt[xxxi]. Amongst a long list of things you would experience immediately would be: fatigue and weakness, difficulty thinking, remembering and reasoning, poor eyesight and digestion. Longer term, more severe effects include poor wound healing, chronic illnesses, allergies and asthma, diabetes and obesity, and finally, heart disease and certain cancers.

About 3 billion years ago, something similar to modern photosynthesis developed. This made the sun's energy available not only to algae, seaweeds and other plants, but also to the plant eating species that consumed them. Photosynthesis used the plentiful carbon dioxide and water as raw materials and, with the energy of sunlight, produced energy-rich carbohydrates. Mitochondria are essential to plants as well. In addition to the chloroplasts that fix solar energy to create glucose as plant food, the mitochondria burn this glucose with oxygen, thereby providing the energy plants need to survive and grow.

When plant respiration and animal respiration and digestion are balanced, enough oxygen is produced to balance out the carbon dioxide emitted by people and other animals. Oxygen is a crucial element for all living creatures, which need it in stable concentrations to survive at all. Curiously enough, the level of oxygen in the earth's atmosphere has hovered at around 21%[xxxii] for a few billion years, with some remarkable exceptions.

Meantime, the levels of carbon dioxide in our atmosphere have also remained fairly steady, at around 0.039 % of our atmosphere—less than half a part per thousand—a tiny amount.[xxxiii] Yet this teensy weensy amount of gas is the reason life on earth can exist at all. Today, the excess carbon dioxide produced by the industrial age over the past 500 years, has unbalanced this natural cycle. In his most recent book, climate scientist James Lovelock[xxxiv] presents evidence that sea levels are rising faster, and Arctic ice is melting faster, than current climate models predict. Lovelock suggests that we may already be beyond the tipping point of our climate being pushed into a permanently hot state.

Under such conditions, Lovelock expects human civilization will be hard pressed to survive with temperatures similar to the Paleocene-Eocene Thermal Maximum when atmospheric concentration of CO_2 was 450 ppm[xxxv]. At that point the Arctic Ocean was 23 °C and had crocodiles in it, with the rest of the world being mostly scrub and desert[xxxvi]. (I am personally less pessimistic, and you will find out why in later chapters. Evolution, thankfully, offers us choices as conscious creatures—and we still have some time to make better, more sustainable choices.)

Too *much* oxygen, however, can also be dangerous. The 'dark side' of photosynthesis is that it produces oxygen as a waste product. In the very early days of life on earth, this free oxygen became bound up with limestone, iron, and other minerals. There is substantial proof of this in iron-oxide rich layers in geological strata that correspond with this time period. We know it as "rust", and this rust-red colour permeates the iron rich rocks of the mountains and coastlines of Catalonia, giving it the name "The Vermillion Coast".

The oceans would have turned green while oxygen was reacting with minerals. When the reactions stopped, oxygen could finally enter the atmosphere. Though each single celled organism only produced a minute amount of oxygen, the combined metabolism of many cells over a vast period of time transformed the Earth's atmosphere. Some of the oxygen was stimulated by incoming ultraviolet radiation to form ozone, which collected in a layer near the upper part of the atmosphere. The ozone layer absorbed, and still absorbs, a significant amount of the deadly ultraviolet radiation that once had passed through the atmosphere. It allowed cells to colonize the surface of the ocean and ultimately the land.

Besides making large amounts of energy available to life-forms and blocking ultraviolet radiation, the effects of photosynthesis had a third, major, and world-changing impact. Oxygen was toxic at even low levels during a period known as the "Oxygen Catastrophe", in which much of life on Earth was wiped out as oxygen levels rose to unprecedented highs. The earth's oceans froze over about 2.4 billion years ago in the Great Oxidation Event. The rising oxygen levels wiped out a huge portion of the Earth's anaerobic inhabitants at the time. From their perspective, it was a catastrophe, though what was bad for Cyanobacteria, turned out to be good for life as we know it today.

This mega-disaster for most single celled organisms was just the moment mitochondria were waiting for. After a suitably chilly and very long period of time, the greenhouse gases emitted by volcanoes are believed to have warmed the earth up again, the oceans melted and then, one sunny day, a mitochondrial cell entered a larger prokaryotic cell. Using oxygen, it was able to metabolize the larger cell's waste products and derive more energy. Some of this surplus energy was returned to the host. The smaller cell replicated inside the larger one, and soon a stable symbiotic relationship developed.

The presence of oxygen provided life with new opportunities. Aerobic metabolism is more efficient than anaerobic pathways, and the presence of oxygen undoubtedly created new possibilities for life to explore. Since the beginning of the Cambrian period, oxygen levels have fluctuated between 15% and 30% of atmospheric volume. Toward the end of the Carboniferous period (about 300 million years ago) atmospheric oxygen levels reached a maximum of 35% by volume, which may have contributed to the large size of insects and amphibians at this time. (And you think mosquitoes are annoying today—imagine a dragonfly sized mosquito hovering over your bed for unprecedented terror!)

Symbiosis developed between the larger cell and the population of smaller cells inside it: they joined to become a single organism with what a hot cellular marketing team might call "Mitochondria Inside"[xxxvii]. In this way what we know today as animals, emerged. A similar event took place with photosynthetic cyanobacteria, which entered larger heterotrophic cells and becoming chloroplasts. As a result of these changes, a line of cells capable of photosynthesis split off from the other multi-cellular organisms (called eukaryotes) around one billion years ago. Thus emerged the first plants.

By around 900 million years ago true multi-cellularity had evolved in both plants and animals. The first plants were in fact so successful that they absorbed almost all of the carbon dioxide needed to keep earth warm enough for life, leading to a very severe ice age around 770 million years ago—so severe, in fact, that the surface of all the oceans completely froze over. This period was known as "Snowball Earth". Ironically we now face the opposite problem—humans have been so successful in producing carbon dioxide that we face major climate change this century, and 25% of the species on earth are now extinct as a result of human activities. We have ourselves become a force of nature, and hopefully with our large brains and big hearts we can escape the fate of the vegetable kingdom i.e. total extinction, during Snowball earth.

Eventually, after 20 million years, enough carbon dioxide escaped through volcanic outgassing; the resulting greenhouse effect raised global temperatures sufficiently so that fish, (the earliest vertebrates) could evolve in the oceans of 530 million years ago, leading to the pre-Cambrian explosion of water-borne life forms. A third major extinction event occurred near the end of the Cambrian period, which ended 488 million years ago.

The communication of information among early life forms was largely automatic. The consciousness of plants and primitive animals received the information it needed from its world largely by instinct. But as consciousness evolved to encompass new levels of depth, new types of communication systems appeared that serve to share and exchange this emerging depth of consciousness. In the time line of evolution, the appearance of new communication systems is evidenced by the development of specialized sounds or movements in animals.

One of the most interesting characteristics of living creatures is their self-organizing property—their capacity for what biologists call autopoiesis (literally, "self-production"). Part of what defines a dynamic system, whether a galaxy, cell, or dolphin, is that it contains within itself the code of its external form and inner structure. Dynamic systems arise from within themselves. And these self-preserving, coherent patterns of autopoiesis are maintained by dynamic systems within an environment of constant change. The technique these dynamic natural systems use to maintain and develop their ordering pattern is through the exchange or metabolism of different forms of energy.

Self-organizing processes continuously create more organization. What sustains the organization in these processes is the energy flowing through the process. The process extracts order from the energy, and in so doing degrades the energy (or in other words, increases its entropy, the measure of disorder). For example, we are all familiar with the way living things use energy to maintain their bodies, and how their metabolization of various forms of energy results in byproducts such as carbon dioxide or feces. And we can see how a life form's relationship to the "food chain" of energy nourishment in which it participates serves to define almost every aspect of its form. For instance, how an animal gathers and uses food (and usually how it keeps from becoming food) determines the structure of its body.

Life on Land Emerges

A food web (or food cycle) depicts feeding connections (who eats whom) in an ecological community—*Wikipedia*

Several hundred million years ago, plants (probably resembling algae) and fungi started growing at the edges of the water, and then out of it. The oldest fossils of land fungi and plants date to 480-460 million years ago, though molecular evidence suggests the fungi may have colonized the land as early as 1 000 million years ago and the plants 700 million years ago. You may notice them gathering today in the nooks and crannies of your shower (mainly black algae), or in your pond or swimming pool (mainly green algae).

Initially remaining close to the water's edge, mutations and variations resulted in further colonization of this new environment. The timing of the first animals to leave the oceans is not precisely known: the oldest clear evidence is of arthropods on land around 450 million years ago, perhaps thriving and becoming better adapted due to the vast food source provided by the terrestrial plants.

Then, 360 million years ago, amphibians emerged onto land and plants evolved seeds enabling them to spread over the surface of the earth, at the same time as another global cooling resulted in a mass extinction. 300 million years ago, the supercontinent called Pangaea formed,. The most severe extinction event to date took place 250 million years ago, at the boundary of the Permian and Triassic periods; 95% of life on Earth died out, possibly due to the "Siberian Traps" volcanic event.

The largest eruption of the 20th century, Mount Pinatubo, is tiny compared with the Siberian Traps. Yet even Pinatubo caused a 0.5 degree drop in global temperatures the year after it erupted. The largest eruption in historic memory occurred on Iceland in 1783-84 spewing out 12 cubic km of lava onto the island (the Siberian Traps erupted about 3 million cu km). The poisonous gases given out are recorded as killing most of the islands crops and foliage and lowering global temperatures by about 1 degree. If events this size can affect temperatures and large areas then the effects of a large scale flood basalt would have been catastrophic.[xxxviii]

Today the giant Norilsk-Talnakh nickel-copper-palladium deposit formed within the magma conduits in the main part of the Siberian Traps is run by some of the world's richest men, proving that for almost every catastrophe there can be a silver lining, eventually.

Then 65 million years ago the next major catastrophe struck. A 10-kilometer wide meteorite ploughed into the Earth just off the Yucatán Peninsula, ejecting vast quantities of particulate matter and vapor into the air that occluded sunlight, inhibiting photosynthesis. Most large animals, including the non-avian dinosaurs, became extinct.

Shortly thereafter (i.e. tens of millions of years in geological time, which is a little slower than our human time)[xxxix], mammals rapidly diversified, grew larger, and became the dominant vertebrates. Perhaps a couple of million years later (around 63 million years ago), the last common ancestor of the primates lived. By the late Eocene epoch, 34 million years ago, some terrestrial mammals had returned to the oceans to become animals such as Basilosaurus which later gave rise to dolphins and whales.

From this brief summary of the evolution of life, we can draw several conclusions regarding the Trouble with Paradise. Firstly, Mother Nature can be a real bitch! Five mass extinctions in the fossil record, and a possible sixth mass extinction on its way. Secondly, despite her historical bitchiness, Mother Nature also has a big upside: she is an open system, which means we can tinker with her bits and pieces to come up with new and better ways of doing things which would previously have been regarded as impossible. Although today we may laugh at the religions of yesteryear from our privileged scientific perspective built up during the past five hundred years of enquiry, we can also appreciate the role that religion has played in creating human cultures and advancing human consciousness over time.

Thirdly, Mother Nature is highly unpredictable. Pilots have a saying: Flying involves hours of boredom, punctuated by brief moments of stark terror. In evolution, we have the "punctuated equilibrium" theory, which states something similar:[xl] instead of a slow, continuous progression, the evolution of life on Earth seems more like the life of a soldier: long periods of boredom interrupted by rare moments of terror. I am sure our ancient ancestors experienced a similar kind of pattern in their hunting and gathering, loving eating and praying.

Volcanoes are living proof that the Earth has been a source of constant inspiration and awe for humanity, motivating a range of creative endeavors. Given their combination of destructive and creative forces, volcanoes have a strong link to religion around the world. Making an offering to a volcano for protection from an eruption is a common practice in Indonesia. At Mt. Bromo, the most worshipped Indonesian volcano, people climb its slopes to pray and give offerings each year on Buddha's birthday. In Japan, however, volcanoes do not have a direct religious role in society, but instead are venerated by the Japanese. Mount Fuji, near Tokyo, is considered a lucky image in dreams.

Residents near a volcano commonly appeal to local religious figures for protection from a volcanic eruption. For instance, on occasions when Mount Vesuvius, in Italy, erupts or threatens to erupt, the residents of nearby Naples bring forth the relics (a skull and a vial of blood) of Saint Januarius, a Catholic Bishop who was martyred around A.D. 300.

Parading the relics through the street and presenting them to the volcano is said to have stopped eruptive activity on several occasions.

Similarly, in Hawai`i, Kamehameha made offerings of breadfruit, fish, and even a pig to Pele in an attempt to stop the destructive eruption of Hualalai Volcano in 1801. Only when the King threw a lock of his hair into the lava, symbolically offering a part of himself, did the eruption stop. Eighty years later, Kamehameha's granddaughter, Princess Ruth Ke`elikolani, who never accepted Christianity, chanted and left offerings to Pele at the front of a lava flow that threatened Hilo during an eruption of Mauna Loa. The eruption soon stopped, although some might argue that Christian prayer meetings being held in Hilo at the time were responsible for saving the city.

Volcanoes also play a prominent role in Christianity. In the year 1000 in Iceland, there was bitter debate over whether the country would become Christian or continue to worship Norse gods. During one parliamentary session on the subject, a messenger arrived with news of a volcanic eruption occurring near the city of Reykjavik. Believers in the old religion took that as a sign of anger from their gods. This prompted a Christian leader to ask what might have angered the gods during previous eruptions which formed the lava plains on which they stood. This question ended the debate, and the nation converted to Christianity.

Here in Catalonia, we have Mont Canigou, part of the volcanic Pyrenean range. The top of Canigou looks as if it has been blown off in some giant explosion, so sharp are its edges. In fact, climbing to the top along a winding footpath, it looks as if a giant has thrown sharp, large, pointed boulders at random across the top of the mountain. Clearly some gigantic forces were at work in the creation of this truncated caldera. Nearby, in Spanish Catalunya, there are still 40 active volcanoes and various lava flows, in the La Garrotxa Volcanic Zone Natural Park. From the Fluvià river valley to the county's northern limits, the landscape changes dramatically, becoming more dramatic and steeper with an abundance of cliffs and gorges in the Alta Garrotxa.

Canigou has tremendous symbolical significance for the Catalans. On its summit there is a cross that is often decorated with the Catalan flag, and every year on the 23rd of June, the night before St. John's day, there is a ceremony called Flama del Canigó (Canigou Flame), where a fire is lit on the mountaintop. People keep a vigil during the night and take torches lit on that fire in a spectacular torch relay to light bonfires somewhere else. Some estimates conclude that about 30,000 bonfires are lit in this way all over Catalonia on that night.

Spectacular jeep tracks on the north side of the massif lead to the Chalet de Cortalets (at 2150 m.) which is a popular outpost for walkers, and there are two ancient and stunningly beautiful monasteries at the foot of the mountain, St Martin-du-Canigou and Saint-Michel-de-Cuxa. It was in this dramatic corner of the world that some of our earliest human ancestors arrived to make their living.

Out of Africa: The Barbecue Finally Emerges from the Mists of Time

Nothing would be more tiresome than eating and drinking if God had not made them a pleasure as well as a necessity.

Voltaire

"A group of archeologists found the remains of a 7700-year-old aurochs barbecue by a river in the Netherlands. (Aurochs are horned ancestors of the domesticated cow, extinct for 400 years.) Analysis of the remains revealed a complex set of rituals surrounding the caveman meal. The group of hunters that caught and killed the aurochs (using a flint ax) first sucked the fatty marrow from the aurochs' bones, then grilled the beast's ribs at the site of the killing. They brought the rest of the meat-scraped scrupulously from the bones—back to their settlement. The archeologists said that the feast took place 1000 years before agriculture came to the Netherlands; the settlement the hunters came from must have survived exclusively on the food they hunted and gathered."

Huffington Post 2011

A small African ape living around six million years ago was the last animal whose descendants would include both modern humans and their closest relatives, the bonobos, and chimpanzees. Only two branches of its family tree have surviving descendants. Very soon after the split, apes in one branch developed the ability to walk upright. Brain size increased rapidly, and, two million years ago, the very first animals classified "Homo" appeared.

Anatomically modern humans, Homo sapiens, are believed to have originated somewhere around 200,000 years ago in Africa; the oldest fossils date back to around 160,000 years ago. The structure of the recently evolved Homo sapiens brain gave this new species some new capabilities, which when added together, gave them a massive, one might even say an almost unfair, advantage: the ability to shape their environment with technologies which enabled them to survive and thrive in ways their ancestors could barely have imagined.

The ability to control fire led to a dramatic change in the habits of early humans. Making fire to generate heat and light made it possible for people to cook food, increasing the variety and availability of nutrients. The heat produced would also help people stay warm in cold weather, enabling them to live in cooler climates. Fire also kept nocturnal predators at bay. Evidence of cooked food is found from 1.9 million years ago, although fire was probably not used in a controlled fashion until 400,000 years ago. What a pity Tautavel man did not have a technology exchange program with other Homo erectus and Neanderthal hunter gatherer bands—he might still be with us today.

All humans alive today are descended from Mitochondrial Eve, a woman estimated to have lived in Africa some 150,000 years ago. This raises the possibility that the Proto-World language could date to approximately that period. There are also claims of a population bottleneck, notably the Toba catastrophe theory which puts human population some 70,000 years ago was as low as 15,000 or even 2,000 individuals, due to the eruption of a super volcano in Sumatra which created a mini-ice age.

Evidence of the use of both fire and stone tools becomes widespread around 50 to 100 thousand years ago; interestingly, resistance to air pollution started to evolve in human populations at a similar point in time. The use of fire became progressively more sophisticated, with its being used to create charcoal and to control wildlife from tens of thousands of years ago. It is more difficult to establish the origin of language. It is unclear whether Homo erectus could speak or if that capability had not begun until Homo sapiens.

The Trouble with Paradise

One of the many interesting wrinkles in the evolutionary story of our species is that we went from being largely herbivorous, to being staggeringly versatile eaters. Our hunting and foraging ancestors found their food everywhere, in a far greater variety than we could access in a well-stocked supermarket today. Our ancestors were getting their animal protein from such a variety of species that we can scarcely imagine the selection. Along with the enhanced extraction of calories provided by cooking their food over a fire, this vast range of food sources ensured our big-brained ancestors could acquire enough energy to survive and thrive in an extreme range of habitats.

As brain size increased, babies were born sooner, before their heads grew too large to pass through the pelvis. As a result, they exhibited more plasticity, and thus possessed an increased capacity to learn and required a longer period of dependence. Social skills became more complex, language became more advanced, and tools became more elaborate. This contributed to further cooperation and brain development.

Although I am still waiting for the time travel machine I ordered decades ago in my youth (the delivery schedule seems to have been affected by a mysterious "loss of time" that we all experience occasionally), I can do the next best thing right now: jump into my ageing Renault Scenic and travel back six hundred thousand years or so to the caves of Tautavel, about half an hour away where the Fenouilledes sandstone range of compressed ancient seabed, meets the hard granite of the Pyrenees.

Of course, it would be remiss of me not to preface any story about cave men who lived at the dawn of time in this part of the world, without mentioning the nature of the ground they were walking on and the mountains they had to climb. Geologically speaking, the Pyrenees are a 430 kilometer long, intra-continental mountain chain that divides France, Spain, and Andorra.

This ancient mountain chain (much older than the Alps which are only ten to twenty million years old[xli]), has its origins many hundreds of millions of years ago in the Precambrian era[xlii]—in other words, it is one of the oldest mountain ranges on planet earth. The configuration of the Pyrenees today is due to the collision between the micro-continent Iberia and the southwestern promontory of the European Plate (i.e. Southern France). The two continents started approaching each other about 100 million years ago and subsequently collided more violently 55 to 25 million years ago.

The Pyrenees get their name from Pyrene, the virginal daughter of a Mediterranean Gaul King who gave the hero Hercules hospitality during one of his famous labors—to wit, to steal the cattle of Geryon. During a drunken feast Hercules raped Pyrene, she gave birth to a serpent, ran away into the woods and poured out her story to the trees. Sadly, she attracted the attention of the wild beasts instead and was torn to shreds. When Hercules found the girls lacerated remains on his return, the now sober hero tenderly laid her to rest. While so doing, Hercules also demanded that the surrounding geography join in mourning and preserve her name.

History records that: "Struck by Herculean voice, the mountaintops shudder at the ridges; he kept crying out with a sorrowful noise 'Pyrene!' and all the rock-cliffs and wild-beast haunts echo back 'Pyrene!'... The mountains hold on to the wept-over name through the ages."[xliii]

This momentous set of geological and historical forces led to the formation of some pretty amazing landscapes, including the smaller Fenouilledes and Corbieres mountains, where the granite thrust up by the volcanic violence of the meeting of tectonic plates, pushed through the ancient seabed. The tectonic upheaval of the region explains the coexistence of layers from very different geological eras. The majority of soil is a mix of clay and sand with variations according to the local region: red sandstone in Boutenac, stony terraces in Lezignan, grey calcareous clay at Queribus and Servies, schistes in the higher zones of the Corbieres and coral limestone at the edge of the Mediterranean. This also makes for complex and rich wines, but we will get to that most important subject a little later on.

From about 65 million years ago dinosaurs roamed the region, and their fossilized bones and eggs abound.[xliv] Then, around 40 million years ago, the glaciers carved out new valleys, and over 3 million years ago, the first distant relatives of our species arrived in the region. Homo erectus and other pre-sapiens from Africa colonized Europe in several waves. About 650 000 years ago Homo erectus arrived in the Pyrenees-Orientales, living in huts and using sophisticated tools.

During this warm period, saber toothed tigers, hippopotamus, deer, rhinoceros and chamois roamed the plains and mountains of the Roussillon region. One of the best preserved homes of Homo erectus was a cave about 30 km from Perpignan near the charming village of Tautavel. The Tautavel cave[xlv] is cut into a cliff above the River Verdouble. In prehistoric times, this cave would have opened out onto the river, but now stands high above the deep gorge cut by the river over hundreds of thousands of years.

A rather shifty looking character by all accounts, Tautavel man had sunken eyes, large jutting eyebrows and prominent jaws with a meagre chin—not someone you would necessarily buy a used stone axe from. He was 1.65 meters tall (5 feet 5 inches) and weighed about 45 to 55 kg. His cranial capacity was inferior to those of Neanderthal man and Homo sapiens—a mere 1,100 cubic centimeters. It appears that although Tautavel man was incapable of speech on account of his limited brain size and primitive vocal tract, he may possibly have developed an early form of symbolic language. Gestures and crudely vocalized sounds must have been crucial for hunting expeditions, not to mention mating rituals[xlvi]. Perhaps they even had an expression for: "Touch me again yes, right there".

Unlike other Homo erectus sub-species, Tautavel man had not discovered how to use fire for cooking and thus ate his food raw. Today you will find that one of the most popular starters on any restaurant menu in this region is steak tartare (raw mince with tasty spices added, served fresh at your table), which requires a relatively tough digestive system, I can personally attest. With his formidable hunting skills and stone tools Tautavel man must have eaten very well. Some of his prey, slaughtered with projectile spears, weighed 700 kilos or more, and included reindeer, deer, beef, horse, bison, rhinoceros, mouflon (mountain sheep), thar and chamois. These last two are goat-like mountain animals. As evidenced by the presence of roughly sawn human bones in the grotto, Tautavel man probably also practiced cannibalism. Not really someone you would want to invite for a dinner party, unless you planned being on the menu as well.

Then, just when things could not really get any better for Tautavel man, another Ice Age began. 550,000 years ago. In this arctic climate, with a glacial wind lashing the mountains at more than 130 kilometers an hour, Tautavel man had to change his diet to woolly mammoths, reindeer, polar foxes, musk oxen and anything with a warm hide that could keep him from freezing to death. We don't know how exactly when his luck ran out, but it appears that around one hundred thousand years ago, Tautavel man suddenly disappeared.[xlvii] What can be said with some certainty, (and I am sure Stephen Jay Gould would agree), is that at this rather sharp point, Tautavel man's equilibrium had definitely been punctuated.

The Artistic Impulse

Ars gratia artis—"Art for art's sake"

Latin cliché

Art, in its broadest meaning, is the expression of creativity or imagination. The word art comes from the Latin word ars, which, loosely translated, means "arrangement". Art is commonly understood as the act of making works (or artworks) which use the human creative impulse and which have meaning beyond simple description. The term creative arts denotes a collection of disciplines whose principal purpose is the output of material for the viewer or audience to interpret. As such, art may be taken to include forms as diverse as prose writing, poetry, dance, acting or drama, film, music, sculpture, photography, illustration, architecture, collage, painting and fashion. Art may also be understood as relating to creativity, æsthetics and the generation of emotion.

The earliest possible artwork yet discovered, the Venus of Tan-Tan comes from between 500,000 and 300,000 BCE, during the Middle Acheulian period. Discovered in Morocco, it is about 6 centimeters long and resembles a human figurine. Although this Moroccan artifact may have been created by natural geological processes, it appears to exhibit traces of human tool-work and bears evidence of having been painted; "a greasy substance" on the stone's surface has been shown to contain a mixture of iron and manganese termed ochre, and indicates that it was decorated by someone and used as a figurine, regardless of how it may have been formed. The earliest known depictional art is from the Upper Paleolithic period and includes both cave painting (such as the famous paintings at Lascaux), portable art (such as animal carvings and so-called Venus figurines like the Venus of Willendorf), and open air art (such as the Fornols-Haut in France).

The artistic impulse is a human birth right, a trait so ancient, universal and persistent that it is surely innate. But while some researchers have suggested that our artiness arose accidentally, as a by-product of large brains that evolved to solve problems and were easily bored, others argue that the creative drive has all the earmarks of being an adaptation on its own. The making of art consumes enormous amounts of time and resources, an extravagance you wouldn't expect of an evolutionary afterthought. Art also gives us pleasure, and activities that feel good tend to be those that evolution deems too important to leave to chance.

What might that deep-seated purpose of art-making be? Geoffrey Miller and other theorists have proposed that art serves as a sexual display, a means of flaunting one's talented palette of genes. To contemporary Westerners, art may seem detached from the real world, an elite stage on which proud peacocks and designated visionaries may well compete for high stakes. But among traditional cultures and throughout most of human history, art has also been a profoundly communal affair, of harvest dances, religious pageants, quilting bees, the passionate town rivalries that gave us the spires of Chartres, Reims and Amiens.

Art did not arise to spotlight the few, but rather to summon the many to come join the parade. Through singing, dancing, painting, telling fables of neurotic mobsters who visit psychiatrists, and otherwise engaging in "artifying," people can be quickly and ebulliently drawn together, and even strangers persuaded to treat one another as kin. Through the harmonic magic of art, the relative weakness of the individual can be traded up for the strength of the hive, cohered into a social unit ready to take on the world. The only social elixir of comparable strength is religion, another impulse that spans cultures and time.

Recent research suggests that many of the basic phonemes of art, the stylistic conventions and tonal patterns, the mental clay, staples and pauses with which even the loftiest creative works are constructed, can be traced back to the most primal of collusions—the intimate interplay between mother and child. The tightly choreographed rituals that bond mother and child look a lot like the techniques and constructs at the heart of much of our art. These operations of ritualization, the affiliative signals between mother and infant, are aesthetic operations. And aesthetic operations are what artists do. Knowingly or not, when you are choreographing a dance or composing a piece of music, you are formalizing, exaggerating, repeating, manipulating expectation and dynamically varying your theme. You are using the tools that mothers everywhere have used for hundreds of thousands of generations.

The Magical Sound of Music

"One of the truly magical things about music and art in general, is the way it transcends language and cultural barriers and embodies universal truths."—

Claire Johnson

"Music is your own experience, your thoughts, your wisdom. If you don't live it, it won't come out of your horn."

Charlie Parker ♪

It has been suggested that the origin of music stems from naturally occurring sounds and rhythms. Human music may echo these phenomena using patterns, repetition and tonality. In certain cultures, instances of their music imitate natural sounds. In some cases, this feature is related to shamanistic beliefs or practice. It may serve also entertainment or practical functions—for example, in playing games or to lure animals in a hunt.

Aside from bird song, monkeys have been witnessed beating on hollow logs. Although this might serve some territorial purpose, it suggests a degree of creativity and seems to incorporate a call and response dialogue. It is possible that the first musical instrument was the human voice itself, which can make a vast array of sounds, from singing, humming and whistling through to clicking, coughing and yawning. The oldest known Neanderthal hyoid bone with the modern human form has been dated to be 60,000 years old, predating the oldest known bone flute by 10,000 years; but since both artefacts are unique the true chronology may date back much further.

Music can be traced to prior to the Oldowan era of the Paleolithic age, when stone tools first began to be used by hominids. The noises produced by work such as pounding seed and roots into meal is a likely source of rhythm created by early humans. The first rhythm instruments or percussion instruments involved the clapping of hands, stones hit together, or other things that are useful to create rhythm. Examples of Paleolithic objects which are considered unambiguously musical are bone flutes or pipes. The earliest unambiguously musical bone pipe is from Geissenklösterle in Germany, which dates to about 36,000BP.

The world's oldest known song is approximately 3,400 years old and written in Hurrian on a clay tablet found at the site of the city of Ugarit in the early 1950s. Due to the lack of confirmatory material translations of the text differ, although all current interpretations agree that the music is diatonic. On some interpretations the music consists of two melodic lines and utilizes both major and minor thirds, on other interpretations the music consists of one melodic line (is monophonic) with a rhythmic accompaniment.

The First Delphic Hymn is the earliest unambiguous surviving example of notated music from anywhere in the world. In the Aegean Sea, north of Crete lies a group of small islands known as the Cyclades. On one of these islands two marble statues from the late Neolithic culture (2900 BC-2000 BC) were discovered together in a single grave. They depict a standing double flute player and a sitting musician playing a triangular-shaped lyre or harp. The harpist expresses concentration and intense feelings and tilts his head up to the light. The meaning of these and many other figures is not known; perhaps they were used to ward off evil spirits or had religious significance or served as toys or depicted figures from mythology. The discovery of this and similar pieces in the late 19th century had considerable influence on the sculpture of the early 20th century, for example on that by modernists such as Picasso and Modigliani.

For physicists and the wisdom traditions of the world, everything is vibration. From the electrons spinning around the nucleus of an atom, to the planets spinning around suns in the galaxy, everything is in movement. Everything is in vibration. And if it is in vibration, it is putting out a sound. Sound travels as a wave form. First, let us say that one way of measuring sound is how fast or slow this wave form is moving. These waves are measured as cycles and sound is measured in cycles per second. This is called its frequency. Very slow waves make very low sounds. Very fast waves make very high sounds. The lowest note on a piano is about 24 cycles per second. The highest note on a piano is about 4,000 cycles per second.

Human hearing ranges from around 16 to about 16,000 cycles per second. Yet just because we can't hear something does not mean it is not vibrating, nor creating a sound. Dolphins can project and receive information upwards of 180,000 cycles per second, more than 10 times that of humans. And this to them is sound. Simply because you cannot hear an object does not mean it is not vibrating or making a sound.

Every object has a natural vibratory rate. This is called its resonance. One of the basic principles of using sound as a transformative and healing modality is to understand the idea that part of the body is in a state of vibration. Every organ, every bone, every tissue, and every system—all are in a state of vibration. When we are in a state of health, the body puts out an overall harmonic of health. However, when a frequency that is counter to our health sets itself up in some portion of the body, it creates a disharmony that we call disease.

Ever since the 1960s there has been much loose talk about raising vibrations. Over-used though this phrase may be, it should not be dismissed out of hand. For the past twenty years, many scientists around the world, including Valerie Hunt, a professor of kinesiology [the study of human movement], have measured human electromagnetic output under different conditions. Using an electro-myograph, which records the electrical activity of the muscles. Hunt, (like Dr Hiroshi Moto Yama, scientist and Shinto priest), recorded radiations emanating from the body at the sites traditionally associated with the chakras. Through her research she made the startling discovery that certain types of consciousness were related to certain frequencies.

She found that when the focus of a person's consciousness was anchored in the physical world, their energy field registered the frequencies in the range of 250 cps (cycles per second). This is close to the body's own biological frequency. Active psychics and healers, however, registered in a band between 400 and 800 cps. Trance specialists and chanellers registered in a narrow field of 800-900 cps, but from 900 cps onwards Hunt correlated what she termed 'mystical personalities' who had a firm sense of the cosmic interconnections between everything. They were anchored in reality, possessed psychic and healing abilities, were able to enter deep trance states, yet had transcended and unified the separate experiences through a mystic, holistic, metaphysical philosophy.

Another study by Dr. Valerie Hunt at UCLA used more conventional measuring equipment in a study of the chakras and the human energy field. Hunt found regular, high frequency, sine-wave oscillations coming from these points that had never previously been recorded (1978). The normal frequency of brain waves is between 0 and 100 cps (cycles per second), with most Information occurring between 0 and 30 cps.

In comparison, muscle frequency goes up to about 225 cps and the heart goes up to about 250 cps. The readings from the chakras was in a band between 100 and 1600 cps—far higher than what has been traditionally found radiating from the human body. Each colour of the human aura is associated with a different wave pattern, and these wave patterns were recorded by Dr. Hunt at the chakra points. When colours such as "white light" were seen in the auric field, the frequency signal measured was over 1000 cps. Hunt has hypothesized that this high frequency level is actually a sub-harmonic of an original frequency signal which is in the range of many thousands of cycles per second: a sub-harmonic of the original subtle energy of the chakra.

Music appears to be a particularly powerful means of eliciting deep spiritual and collectively magical experiences, enabling the hairs on the back of our head and neck to rise in response to the sublime beauty deep inside the music and ourselves. Music is a mysterious force that bonds us with each other in the moment, and dissolves our differences for the short while we are embraced in its arms. Through its ability to forge such powerful bonds, music has played a key role in the evolution of our species, keeping us close in times where the solidarity of our small hunter gatherer groups was a key to our survival.

And as our music has evolved into more sophisticated forms, so too has the technology we use to make it and enjoy it. Developments in music, art and technology have always been closely intertwined, enabling our forms of expression to reflect the increasing sophistication of ourselves and our societies over time.

Dr Robin Lincoln Wood

Tools & Technology

Technology is a broad concept that deals with a species' usage and knowledge of tools and crafts, and how it affects our ability to control and adapt to our environment. In human society, it is a consequence of science and engineering, although several technological advances predate the two concepts. Technology is a term with origins in the Greek "technologia": "techne" ("craft") and "logia" ("saying").

The use of tools by early humans was partly a process of discovery, partly of evolution. Early humans evolved from a race of foraging hominids which were already bipedal, with a brain mass approximately one third that of modern humans. Tool use remained relatively unchanged for most of early human history, but approximately 50,000 years ago, a complex set of behaviors and tool use emerged, believed by many archaeologists to be connected to the emergence of fully-modern language

People's use of technology began with the conversion of natural resources into simple tools. The pre-historical discovery of the ability to control fire increased the available sources of food and the invention of the wheel helped humans in travelling in and controlling their environment. Recent technological developments, including the printing press, the telephone, and the Internet, have lessened physical barriers to communication and allowed humans to interact on a global scale. However, not all technology has been used for peaceful purposes; the development of weapons of ever-increasing destructive power has progressed throughout history, from clubs to nuclear weapons.

Technology has affected society and its surroundings in a number of ways. In many societies, technology has helped develop more advanced economies (including today's global economy) and has allowed the rise of a leisure class. Many technological processes produce unwanted by-products, such as pollution, and deplete natural resources, to the detriment of the Earth and its environment. Various implementations of technology influence the values of a society and new technology often raises new ethical questions. Examples include the rise of the notion of efficiency in terms of human productivity, a term originally applied only to machines, and the challenge of traditional norms.

Philosophical debates have arisen over the present and future use of technology in society, with disagreements over whether technology improves the human condition or worsens it. Neo-Luddism, anarcho-primitivism, and similar movements criticize the pervasiveness of technology in the modern world, claiming that it harms the environment and alienates people; proponents of ideologies such as trans-humanism and techno-progressivism view continued technological progress as beneficial to society and the human condition. Indeed, until recently, it was believed that the development of technology was restricted only to human beings, but recent scientific studies indicate that other primates and certain dolphin communities have developed simple tools and learned to pass their knowledge to other generations.

The earliest modern humans—Cro-Magnons—entered France around 40,000 years ago during a long interglacial period of particularly mild climate, when Europe was relatively warm, and food was plentiful. Actually, coming to think of it, that also probable explains some of the strange driving you see on French roads even today. In fact I think I've even seen a few Cro-Magnon families on the Paris underground and in the Louvre Museum, but that, as they say, is another story.

When they arrived in Europe, they brought with them sculpture, engraving, painting, body ornamentation, music and the painstaking decoration of utilitarian objects. Some of the oldest works of art in the world, such as the cave paintings at Lascaux in southern France, are datable to shortly after this migration.

The first humans to show evidence of spirituality are the Neanderthals—they buried their dead, often apparently with food or tools. However, evidence of more sophisticated beliefs, such as the early Cro-Magnon cave paintings (probably with magical or religious significance) did not appear until some 32,000 years ago. Cro-Magnons also left behind stone figurines such as Venus of Willendorf, probably also signifying religious belief. By 11,000 years ago, Homo sapiens had reached the southern tip of South America, the last of the uninhabited continents. Tool use and language continued to improve while interpersonal relationships became more complex.

The rapid development of our species during the last 40 000 years is attributed by experts to the virtuous circle of co-evolution of our arts, languages, music and technologies together with our need to adapt to very different food sources. During this time our consciousness has evolved beyond what leading psychologists call Type 1 or intuitive awareness and cognition centered in the here and now, to Type 2 consciousness which recognizes past and future, and is able to reflect on and apply learning from experience.

At its most advanced this leads to a number of different kinds of symbolic intelligence and refined pattern recognition which lie at the core of our educational, professional and public lives. Depending on how we classify them, such multiple intelligences can number from a handful to dozens. Each of these more advanced intelligences has played a key role in the evolution of our civilization, and will no doubt be crucial to the evolution of our species in the future.

Culture

"Beauty is eternity, gazing at itself in a mirror. But you are eternity. And you are the mirror."

Kahlil Gibran

Culture is like a giant mirror which enables us to see who we are more clearly. The various facets of a culture also provide us with the means to change what we do not like in the mirror, and retain the things we cherish most. To see ourselves clearly in the mirror of our culture, however, we need to be conscious or "awake" enough to use our reflective, Type 2 intelligences in a discerning way.

In our daily human lives we are each generally in the grip of our own unconsciously held perspectives, cultures and values which are both unseen and taken for granted, much, like a fish takes water for granted. A fish only realizes how crucial water is to it, when it is flapping wildly on the bank of the river or the rocks trying to launch itself back into the water, back into life. Even then, such a discovery is unlikely to result in the concept "water" becoming part of the vocabulary of fish, given their relatively tiny brains.

We have all probably experienced what this is like at some level as we have grappled with the challenge of changes in our own environments, whether voluntary or imposed on us. It may begin with something as simple as experiencing new food, music or cultures in the safety of our existing environment. Then the desire to experience "the real thing" takes hold, and we are off on a journey of discovery. Travellers consciously choose to explore a different city or country, a new language, or perhaps even a new religion.

Many species on earth are social animals, including the ants and bees in the insect kingdom, the dolphins and whales in the sea, certain kinds of birds, the great apes and ourselves. We humans are evolving through our social and cultural media, as we invent and innovate new ways of doing things and new ways of being. This social aspect of our evolution leads us to the second factor in our unconscious evolution: social and political networks.

For millennia we have engaged in behaviors and adaptations which enable us to live together in large social groups such as hunter gatherer bands, tribes, and then nation states. We have not consciously selected these new modes of organization—they have simply evolved as we have out of our need to make these larger social systems work, and to extract enough energy and food to keep our growing populations alive and well.

Throughout most of its history, Homo sapiens lived in small bands as nomadic hunter-gatherers. The transition from the Early to Middle Stone Ages coincides with the emergence of the modern Homo sapiens from earlier, related archaic human species. The next transition from early hunter-gatherer societies to the agrarian and agricultural societies is known as the Neolithic Revolution.

The use of language is one of the most conspicuous traits that distinguish Homo sapiens from other species. Sometime around 50 000 years ago scientists hypothesize that a crucial genetic mutation occurred which enabled anatomically modern humans to develop complex languages. According to the 'Out of Africa' hypothesis, at this time a group of humans left Africa and proceeded to colonize the rest of the world, including Australia and the Americas, which had never been populated by our species before. Some scientists believe that Homo sapiens did not leave Africa before that, because they had not yet attained modern cognition and language, and consequently lacked the skills or the numbers required to migrate.

In the early 21st century, it is difficult not to imagine a globe with human beings spread across its four corners, yet this development of language was the key to our recent and future evolution. Instead of being selected only for physical attributes such as our size, cunning, beauty, speed and strength, language took our species to a new level—we (and nature) could select amongst more abstract characteristics such as ideas, hypotheses, and the beauty of a phrase or story. Modern languages enabled the spread of the cultural artefacts we had started to create with art and music since the dawn of our species.

At the same time, the emergence of inter-subjective relationships became possible due to the emergence of culture in Homo sapiens. Inter-subjective relationships are "patterns in consciousness that are shared by those who are 'in' a particular culture." So it is not the sign or the symbol or the external act or artefact that comprises the fundamental relationships of the inter-subjective world, but the connection or accord between the subjects participating in the communication. That is, inter-subjective relationships exist in the shared connection—the overlap of consciousness that exists between subjects—and so they are neither wholly subjective consciousness nor wholly objective events.[xlviii]

A complex inter-subjective relationship does not arise from symbolic communication until there is a kind of agreement about what the symbols mean. That is, the successful communication of meaning requires interpretation—the transmission of meaning depends on a kind of mutual understanding. For the meaning to be shared there must be an interior harmonic resonance of depth. So, just as the emergence of the cell represents a transcendent breakthrough in the evolution of the biosphere, so too does the emergence of depth-sharing agreements between subjects represent a breakthrough in the inter-subjective world we call our culture.

By "agreement" I mean the receipt or exchange of interior meaning or value, not just the sound. This type of agreement can be formed even if the subjects don't "agree with each other"—as long as they understand what is meant. An agreement is thus the most basic type of inter-subjective relationship that arises as a result of the successful communication of the interior depth of consciousness, and is the basis for all living culture.

From Egocentric to Ethnocentric Cultures

"Travel is fatal to prejudice, bigotry and narrow-mindedness, and many of our people need it sorely. Broad, wholesome, charitable views cannot be acquired by vegetating in one's little corner of earth."

Mark Twain

In the mists of time, some fifty thousand years ago, Homo sapiens roamed the earth in small hunter gatherer bands of between twenty to perhaps a few hundred individuals, made up of extended families. For perhaps months or years at a time, these bands would roam undisturbed through their local grasslands, forests or mountain meadows looking for food and safe places to sleep, raise their children and care for their elders.

Occasionally, and with usually dramatic consequences, one band would encounter another when moving to a new hunting ground. Although it is impossible to say on how many occasions such encounters led to collaboration rather than conflict, the fossil record makes it clear that many of them ended in massacres of one kind or another. But on other occasions, perhaps intelligence and diplomacy of a kind prevailed, and collaboration became possible.

The history of our species has been the gradual triumph of such collaborative endeavours over the intermittent massacres we have carried out on an increasing scale as we have developed more powerful and deadly technologies. Civilization has gradually been built, layer by layer, through our ability to exchange goods and services with each other, and our ability to generate surpluses through our taming of nature.

In Europe, we have something in the region of four hundred different "tribes" remaining today, gathered into 27 nation states inside some very odd borders indeed. In Africa, there are also somewhere in the region of four hundred different "tribes" today, stitched into the dysfunctional borders created by the divide and rule policies of colonialisation. I have not had time to count the number of tribes in the Americas or Asia yet, but my central point is that we have evolved to be strongly connected into and identified with a tribe, comprising not only our extended family but also the cousins of our cousins and beyond. In

China today for example, business trust or "guanxi", extends no further than such extended family ties. This is also the root of corruption in countries where such tribal networks continue to channel power and money toward "Cosa Nostra".

While such tribalism has been essential for our local affairs to work at all, and essential to the regulation of local economies and politics, as our species has expanded into ever more limited space and locales, we have begun bumping up against each other in ways that have resulted in a particularly bloody series of history books for the different cultures and tribes around the world.

While there have been some notable exceptions to this rule, especially in matriarchal societies, the history we are taught today is centered around battles and wars for good reason—they have been a central pre-occupation of our species for millennia, and given the hundreds of millions of people who died in wars in the 20th century alone, appear to be an ongoing certainty, though perhaps on a smaller scale as we slowly knit together our planet into a single, globalized system.

From Ethnocentric to Worldcentric Cultures

"Your car is Japanese. Your pizza is Italian. Your falafel is Lebanese. Your democracy is Greek. Your coffee is Brazilian. Your movies are American. Your tea is Sri Lankan. Your shirt is Indian. Your oil is Saudi Arabian. Your electronics are Chinese. Your numbers are Arabic, your letters Latin. And you complain that your neighbor is an immigrant? Pull yourself together!"

Anonymous

Our bodies grow and develop through a process intelligence which involves the balancing of different systems in the body and environment contending for attention and resources, such that the whole body is optimized and coherent. This involves incredible amounts of communication and bandwidth at the speed of light—much like the balancing going on globally between contending cultures and entities, any one of which could destroy our species if it were to become too dominant.

As we evolve toward world-centric consciousness and systems, the most important and potentially dangerous interactions are those between the leaders of our nation states and within those nation states, the political networks which determine how power is distributed over time. Again, this has up until now been a largely unconscious process as we have not had any means of viewing how this process functions until the advent of mass communications including newspapers, the radio, TV and the internet. Recent moves toward government transparency in the UK, USA and S Africa have now begun a trend similar to that in the trend toward transparency in corporate reporting, ensuring that government is really by the people, of the people and for the people and not just for the politicians and their corporate cronies, at least some of the time!

Probably the most important development today is the emergence of the awareness of the need for some form of global governance by the leaders of the world's most powerful nations. The events set in motion in April 2009 are a good example, where the 20 leaders of the G20 nations representing 80% of the world's economic might gathered to discuss how to reform the global financial and monetary system. This marked the beginning of an ongoing process in which global issues such as the global green new deal to deal with climate change, and a global deal on developing a proto-type system of governance for the planet.

Social and political networks co-evolve as a result of the changing distribution of power within large organizations and governments. Equally important today is the role of international non-governmental organizations (INGO's) and not-for-profit organizations and NGO's who shape the policies and strategies of governments and organizations based on the specific issues they were set up to deal with. For example, INGO's that currently wield a lot of power and influence include the United Nations and all its programs and departments, the European Commission and all its programs and departments, the World Bank, the International Monetary Fund, the G7 and G20 groups of governments leaders, and the regional groups such as ASEAN, the African Union and many others.

All of these institutions are a good start toward a global system of governance, yet their very success depends on the ability of nation states to align themselves around common goals and principles. A global system of *governance* should be contrasted with a global *government*. If the latter ever happened, it would be a risky undertaking indeed due to the sheer scale of the potential for the abuse of power and lack of accountability. The tendency of our time for ethnocentric alpha males to concentrate and abuse power is such that even national federal systems such as exist in varying degrees in the USA and Europe, are subject to frequent abuse.

In contrast, a global system of governance would contain very strong checks and balances to ensure that the will of the people was always paramount, and that bottom-up institutions representing the true diversity of humankind, thrived. Through online campaigning organisations such as Avaaz and 350.org, we can already see the power of such bottom up representation to shift major decisions in the parliaments of the world. And we have only just begun to see the power that we can wield through even better forms of organisation and campaigning in the future. As human consciousness shifts from an ethnocentric to a world-centric level in the next few decades, I believe we will see the emergence of many of the components of a functioning global system of governance.

Inter-subjective consciousness has not only played a key role in the evolution of our species, but is now also modifying the evolution of our entire planet. Individual or "personal" self-reflective consciousness is now evolving into collective consciousness as we move from egocentric and ethnocentric modes of being to a world-centric perspective. We do, however, need to be patient regarding the speed of this next step for our species. It will take at least another century for worldcentric consciousness to permeate our planet, but that is a mere blink in the evolution of Homo sapiens.

Chapter Five
Paradise Began in a Garden

At a practical level, then, what does this all mean in daily life? While we can appreciate the mega trends underlying the major changes and evolution we see going on in the world around us, what does this mean for each of us, at a personal level? And how would that scale from us personally, to our families, communities, cities, organizations, countries, and "the planet"?

On moving from London to France, I bought a quaint coffee mug illustrated with Victorian style garden implements and vegetables surrounding the words: "Head Gardener". I bought the mug at a time I was beginning a project I dubbed "Oasis", which began with me buying a derelict Chateau in Perpignan between the Pyrenees and the Mediterranean, and then spending several years renovating it. On a good day I start my morning with my Head Gardener mug, and some fine Ethiopian fair trade beans, ending up with a frothy, buzz inducing, cappuccino.

Taking on the role of head gardener at the Oasis was an opportunity for me to try out a hunch I had been working on since I was a tree-planting teenage Junior Mayor of the small, countrified town of Sandton in South Africa. As a youngster, the beautiful gardens planted and tended by my mother and father on the weekends, in Johannesburg, Toronto and Sandton, led me to be a keen nature lover and proud gardener from an early age. Gardening is therapy for the many stuck in the modern rat race, and certainly a much more productive way of spending an afternoon than smacking a little white ball around a golf course.

My hunch was that the disconnect between ourselves, nature and each other proliferated by the machine model of the world perfected during the Industrial Revolution, might be healed if we not only did a little gardening ourselves, but also thought like head gardeners when designing cities, buildings, organizations, products and services.

I started studying management, economics, law and languages during my first year at University. Over the course of the next six years

I explored different models of management, production, marketing, labour, economics and finance, while also becoming an Advocate of the Supreme Court and working as an accountant in my hoidays. The models hidden beneath the surface of each of these disciplines all suffered from physics envy, in their attempts to reduce everything to a mathematical equations and numbers. Without realizing it, we were all learning to design and manage well-oiled machines.

At the same time I became deeply involved in the anti-apartheid struggle and the environmental movements, which taught me an entirely different lesson. Our political theories and practice seemed to be about domination of one ideology over another, with little regard for the human beings in between. And I also learned that politics is a barely disguised dominance game played in islands of hierarchical order in power structures emerging from the chaos of suppressed human energy, aspiration and frustration. It was here I learned the basics of leadership.

As I struggled to reconcile the contradictions between the chaos of human affairs, the horrendous side-effects of our industrial machine, and the beautifully emergent order in nature, the place I felt most at peace was in nature. It is only now, forty years later, that I have begun to see the outlines of the principles by which a master gardener would design and build a better model for our civilization than the deadly, cold machine we created between 1850 and 1950, and that continues to drive globalization around our planet today.

Why We Need Gardens More Than We Know

Our species is one of hundreds of millions on our planet. Like all the rest, we evolved over billions of years, but unlike the rest, we also evolved to have one of the largest brains relative to our body weight (for a land-based creature). We developed opposable thumbs, and then began to develop technology. Yet until recently, we have always lived as part of nature, taming the wilder parts of our planet to meet our needs for food, shelter and belonging. What we today call gardens, were originally the places where we tamed parts of the wilderness to more easily produce our food within easy range of our shelters.

Having learned how to irrigate fields, cross-breed more productive strains of seeds and animals, we were able to produce surpluses for the first time in our evolution. The first human cities emerged, along with the trade routes that enabled us to trade our surpluses with each other. Elites with leisure time emerged, began to stare into still ponds and ask themselves: "Who am I, and How did I get here?" They had banged up against the very first principle of living systems: "*Self-creation*".

Unable to explain the spontaneous emergence of life from clay minerals in the ancient seas of billions of years ago as scientists can today, they invented a useful series of increasingly inventive myths, with gods plural or God singular, being our creators, and indeed the creators of our worlds. So began religion, priesthoods and state power with Kings and Queens acting as the representatives of these divine beings.

In a garden, every tree, shrub, flower, blade of grass, bird, worm, lizard and all the rest are actually wild species that happen to find the way we protect ourselves and our gardens from harsh conditions, convenient. Although we have domesticated dogs, cats, horses and other pets, they sadly make it very difficult to experience our wild guests au naturel.

In our gardens we can also see other principles of living systems in action—for example, there are tens of thousands of different parts, from the biggest trees to the tiniest of mites living in the soil recycling the dead leaves and cellulose. This *complexity contains a great diversity of parts*, and is organized so that the *smaller parts are either embedded in or dependent upon the larger parts*, while the larger parts are also dependent upon all of the other parts too. Thus we have our second and third principles of living systems, *complexity* and *interdependence of parts*.

Within each of the plants and animals in our gardens, we also find that they are both *self-regulating* and *self-reflexive*. That is, they maintain themselves and communicate within themselves so that each part is aware of the state of the other parts. The fourth and fifth principles of living systems work closely with the ninth to fifteenth principles, in that this *communication between all the parts* enables them to *coordinate their functions*, so that *each part is fully empowered to fulfill its specialized role* to the best of its ability. And *if one part is struggling to do so, the other parts will offer assistance* to the struggling part to give it an opportunity to heal.

Not only have living systems negotiated a *healthy balance of interests among their parts, the whole organism and the bigger systems in which they are embedded*; they can maintain this balance across a very wide range of extreme conditions, for very long periods of time, and *evolve new coping mechanisms* as required.

Of course, not only do "Rough winds shake the darling buds of May" as Shakespeare put it 400 years ago—we also find that adverse events from ice ages to meteorites can pose serious threats to all living systems in particular locations. When everything changes, then the sixth, fourteenth and fifteenth principles of living systems snap into action. First of all, *all living systems can respond to internal or external stresses by adjusting the way in which they operate individually*, as well as by *coming together in social arrangements that enable them to radically improve their effectiveness* at finding food and staying alive.

If a major change in the environment persists for long enough, then only the creatures that are better adapted to the new conditions will survive, and they will spontaneously evolve new capabilities. Charles Darwin discovered this more than 150 years ago, and he called it evolution. *Evolution conserves what works well, and creatively changes what does not work well*, all by itself, thus demonstrating the fourteenth and fifteenth principles of living systems. The birds we hear singing in our gardens were once dinosaurs, while the worms ploughing their way through the soil have evolved very little in the past few hundred million years.

Finally, all parts of nature, gardens being no exception, *transform inputs of matter and energy into various kinds of outputs*, while *exchanging matter, energy and information with other parts*. What we see when we look out over the green landscapes we love to nurture, maintain and frolic in, are the stocks of matter, energy and information embodied in the trees, shrubs, flowers, animals and birds. What we do not see, are the flows of energy, matter and information hidden from our eyes between all the different parts of the system, taking place 24 hours a day, 365 days a year. We might spot a lizard's tongue flicking at a fly out of the corner of our eye occasionally, or a bird catching a warm, but most of the time nature does its work silently and out of sight.

In autumn we can see the leaves fall from the trees, while in spring we watch the darling buds emerging miraculously from the stems of trees and flowers, while the birds chirp and dart hither and thither building their nests and attracting mates. The seasons provide us with a wonderful opportunity to see just how creative and adaptive living systems really are.

The 16 Main Principles of Healthy Living Systems

1. Self-creation (autopoiesis)
2. Complexity (diversity of parts)
3. Embeddedness in larger holons and dependence on them (holarchy)
4. Self-reflexivity (autognosis/self-knowledge)
5. Self-regulation/maintenance (autonomics)
6. Response ability to internal and external stress or other change
7. Input/output exchange of matter/energy/information with other holons
8. Transformation of matter/energy/information
9. Empowerment/employment of all component parts
10. Communication among all parts
11. Coordination of parts and functions
12. Balance of Interests negotiated among parts, whole, and embedding holarchy
13. Reciprocity of parts in mutual contribution and assistance
14. Efficiency balanced by Resilience
15. Conservation of what works well
16. Creative change of what does not work well

Thanks to my friend and renowned evolutionary biologist Elisabet Sahtouris for this excellent summary. You can find more information about her work and book "Earthdance" here: *http://www.sahtouris.com/#8_1,0,*

What If We Managed our Human Systems Like Gardens and Nature

So, if nature is so perfect, you might ask, why does life appear to be such a messy and chaotic process? Well, it is absolutely true that the earth's crust, our weather systems, and wilderness in general, are chaotic systems. The result of the massive flow of energy from our sun hitting the surface of our planet every day, is an imbalance between the heat at the surface of the planet and the frozen vacuum of outer space. This imbalance drives our weather, our ocean currents and all living systems.

While the earth's crust and our atmosphere are driven by the basic principles of physics and electromagnetic fields (with plenty of chaotic systems behavior), our biosphere is the product of emergence of living systems that can be partially described by chemistry and biology. The unique feature of our species is, however, more symbolic than physical—we are one of the few species that uses sounds and gestures to communicate, and the only species that has developed written and spoken languages.

This language instinct has enabled us to develop storytelling and the arts to a very high level, as well as to be able to store vast amounts of information both within our minds and in our technologies and infrastructures.

All living systems are also by definition anticipatory systems, in that they have learned at some level how to respond to a range of conditions that enable them to adapt when new situations arise. This is done mainly through physical adaptation, as well as instinctive and conditioned learning.

While most of the living systems on our planet appear to be living mainly in the Now, Homo sapiens has developed a well-refined sense of past, present and future. Our retrospective and prospective memories—memories of the past and memories of the future, and the ability to store and share stories and artifacts about the past and future enable us to develop hypotheses and theories about our past and future, as well as to expand our knowledge and control over our life conditions. Yet with this capability we are equally good at creating as many mini-hells on earth, as we are at creating small corners of paradise.

Outside-In Evolution

Our lives can evolve as a result of the pressure of external forces and life conditions, so that we find our destiny determined by the dominance of the external world over our inner life. For many in the developing world this is the case: without the resources and education the lives of those in poverty usually limit their options for personal development and conscious evolution, with some rare exceptions.

In the developed parts of the world, those with an obsessive attachment to "progress" and "success" often become victims of their obsessions. The happiness of such obsessives then depends on the beneficence or harshness of their environment, with their life strategy being reduced to "get rich, be recognized, be successful", no matter how. Equally the fatalists who seek to escape the Darwinian pressures of the rat race in the belief that the entire system is doomed, become victims of their escape, and hence powerless to make a real difference, even as they feel safer and less stressed as a result of playing the victim.

Never before have so many lived in such comfort and security, despite the major disruptions and wars of the past century,. In the developed world, we are the most materially privileged generation in history. Great strides are also being made in eradicating disease and poverty in the developing world, even if the Millennium Development Goals are unlikely to be met.

The problem is that our obsession with control has led us to expect perfection, and granted us the wish to live in paradise. Yet for the most part our artificial urban paradise is completely divorced from, nature, and relies on an unsustainable global production machine that must work 24x7x365 to grant us our every consumer wish. Not only is this completely unsustainable, it also produces a legion of different kinds of hell on earth for the inhabitants of these overcrowded, false expectation machines.

Dr Robin Lincoln Wood

Inside-Out Evolution

"We are at that stage where the real work of humanity begins. This is the time where we partner Creation in the recreation of ourselves, in the restoration of the biosphere, and in the assuming of a culture of kindness where we live daily life in such a way as to be reconnected and charged and intelligenced by the source of our reality so as to become liberated in our inventiveness and engaged in our world and tasks."

Author and Cultural Philosopher, Jean Houston, Personal Communication

At the other extreme, we can seek to drive our personal evolution from a set of goals and desiderata which emerge from our own personal performance measures, which can often be summed up as exhortations to be or become better in oneself one way or the other. Or perhaps we are a perennial seeker after "truth", believing that if we can just find the right knowledge, all will be explained and our lives will be more meaningful and satisfying. History is full of exceptional individuals who held themselves up to very high standards, and who perhaps were much more evolved beings than our selves: Buddha, Jesus, Mohammed, a Luther, a Gandhi, a Mandela, a Martin Luther King, a Barack Obama.

Such evolved beings were able to change the world around them by being the change they sought in the world. It is a myth to believe that only such exalted figures are capable of doing this—we are all, at our own level, able to do so. The key is to be able to find a niche in the rich evolving pageant of life which enables one to become the change one seeks in the world.

So, where to begin? In my own experience it is useful to recognize that both outside-in and inside-out evolutionary forces are at work around and in us almost all the time. The only question is to what extent the one prevails over the other at any moment, and how one should deal with that in that moment. Just as involution and evolution require a balance in the self and the collective, so too does the process of alignment.

Resonance enables us to appreciate the different forces at work both from within ourselves and from without, while alignments can occur both spontaneously or deliberately. For conscious evolution to become the driving force in our lives and our collective arrangements, we need to align ourselves in contexts and life conditions with others who are travelling in a similar direction to us—otherwise we simply become an unintended victim or temporarily lucky recipient of the flow of unconscious evolution around us, depending on circumstances, and whether we are in the right or the wrong place at the right or the wrong time!

The arc of our civilization and our evolution has slowly been turning from a preoccupation with exploration and conquest (whether military, economic or cultural), to a focus on interiors, both physically and spiritually. At the physical level, particle physics, nanotechnology, genetics, cell biology, information technology, to name just a few disciplines, are taking us ever deeper into the substance of our material world, and enabling us to transform it in clever new ways. At the spiritual level, we find philosophers and cultural leaders exploring and mapping the clash and fusion of all the cultures and ideas that have ever arisen on the planet, while developing new ways of developing ourselves as post-modern avatars of enlightenment.

Despite the great progress in the past fifty years transforming business and government in some places, they remain monolithic modern institutions that privilege themselves over everything and everyone else, treating most people as mere ciphers. We remain in the grip of unconscious metaphors of managing the economy as a machine: (*priming the pump, accelerating demand*), while businesses are still expected to *deliver value* and *conquer new markets* as if there were new worlds to be tamed by our modernist executives beyond our already overcrowded planet.

Yet over the past four decades the culturally creative baby boomers and then the millennial generation have become increasingly sensitized to the alienation, manipulation, psychic deprivation and environmental damage that an obsession with economic growth at any cost is causing. They are reacting, generally peacefully, sometimes violently, against this non-sense, as we can see in the Occupy and other movements, as well as the Arab Spring and similar movements around the world.

This shift in consciousness from modernism at any price to a post-modern and even post-post-modern (or integral) level is changing the game and creating a new playing field for business, government, non-governmental organizations and social entrepreneurs.

The 88 000 multinational corporations, 30 000 NGO's, 195 national governments, hundreds of thousands of local governments and businesses need to be equipped to understand these shifts in order to evolve themselves to meet the demands of 21st century citizens, consumers and culturally creative, inner-directed cybernauts.

They need to begin this journey by undertaking a journey of exploration into their own interiors, learning to go far deeper than the somewhat superficial and tentative attempts of the past few decades to articulate corporate, civic and national values, cultures and ethics. Such attempts have merely scratched the surface of the profound depths that await the organizational explorers and transformers of the future. Such explorers need a sturdy set of principles to guide them, and I believe we can find them by thinking about our social systems, organizations and infrastructures as living systems, closely related to the gardens we are all so familiar with.

When alignment occurs in a personal field, individual coherence becomes evident in the aligned being. When alignment occurs in a collective field, the members of the aligned group may also sense the possibility for coherence in that group and around the group. Coherence is to life what super-conductivity is to matter—it enables energy to flow effortlessly and at or near light speed. The flow of conscious evolution is dramatically accelerated when we are coherent together. Evolving our emerging planetary civilization into a thriving future may depend on just that.

Chapter Six
On the Origins of Heaven and Hell

"Processes of evolution, including variation and natural selection, niche creation and co-evolution, even catastrophe and fluctuating rates of evolutionary change, suggest that adaptation is usually imperfect, with abundant glitches that, as long as they don't constitute abject failures, usually continue to exist unless selection and variation conspire to find a way to get rid of them."

Neuroanthropologist Greg Downey

Human suffering has been endemic and very well documented over the ages, and there are an almost infinite variety of things that can turn a heavenly day into a nightmare, in the twinkling of an eye. Popular culture in the early 21st century also appears to be conditioning people to expect the worst, to a greater extent than was the case in slightly more optimistic decades in the western world, such as the 1950's, 1960's and 1990's. There also appear to be many real causes of suffering in the early 21st century, which are amplified by a bad news seeking media industry and an increasingly sensitive, urban population that feels better watching others suffer from a distance, hoping at the same time that they will not meet a similar fate.

Recently the "The Scream" by Edvard Munch became the most expensive painting in the world upon its sale at a Sotheby's New York auction. Wall Street financier Leon Black paid USD 120 million for this depiction of the modern condition. Everything about the painting is disturbing, from the mustard and orange turbulent sky, the ghoulish face through which a scream ripples outward from the open mouth, the zombie like yellow faced people on their way home from work, and the eerie ship on the fjord that looks like Dracula's ship coming to bring death. Was the Wall Street financier trying to say something about his world, or the world in general? And where exactly would one hang a painting such as the Scream—in the entrance hall to greet one's guests?

Fear, uncertainty and doubt have been with our species for a very long time, but in the past century it appears that anxiety levels have risen to unprecedented highs. It is undeniable that ours is an age in which an enormous and growing number of people suffer from anxiety. The National Institute of Mental Health in the USA tells us that, anxiety disorders now affect 18 percent of the adult population of the United States, or about 40 million people.

In comparison, mood disorders—depression and bipolar illness, primarily—affect 9.5 percent. That makes anxiety the most common psychiatric complaint by a wide margin, and one for which westerners are increasingly well-medicated. Anti-anxiety drugs are being prescribed in unprecedented amounts, resulting in hundreds of millions of prescriptions each year. But are Americans and the rest of us more anxious than our ancestors, or are we simply better diagnosed and treated, while more aware of our mind's tendency to zoom in on our unmet expectations and catastrophize about possible negative events in the future?

Harvard psychologist Robert Kegan believes that we are facing a mismatch between our mental capacities and the complex demands of modern life. As parents and partners, employees and bosses, citizens and leaders, we constantly confront a bewildering array of expectations, prescriptions, claims, and demands, as well as an equally confusing assortment of expert opinions that tell us what each of these roles entails. Kegan brings together the disparate expert "literatures," which normally take no account of each other, to reveal what these many demands have in common. He shows us that our frequent frustration in trying to meet such complex and often conflicting claims results from a mismatch between the way we ordinarily know the world and the way we are expected to understand it through the means of our technocratic culture.

In an age in which Kegan and others tell us that we are "In Over Our Heads", might there be other periods in history when the more one knew, the less sense the world made? In the "Dark Ages" of the fourteenth-century in pre-Renaissance Europe, the Black Death wiped out as much as half the population in four years, along with devastating famines, peasant revolts, religious turmoil, and hordes of pillaging mercenaries. This resulted in waves of mass anxiety attacks. Nor did the omnipresence of the Catholic Church help matters. In fact, it probably

The Trouble with Paradise

made things worse. A firm belief in God and heaven was near-universal, but so was a strong belief in the Devil and hell. And you could never be certain which way you were going to go!

That is pretty hellish by any standard, but today we have developed self-awareness to a point where we've become fixated on our anxious condition. Anxiety didn't emerge as a psychiatric concept until the early 20th century, when Freud highlighted it as "the nodal point at which the most various and important questions converge, a riddle whose solution would be bound to throw a flood of light upon our whole mental existence."

In exploring the origins of heaven and hell, it thus makes eminent sense to begin with hell. From our journey together so far, the Trouble with Paradise turns out to be a feature of several different levels of our existence: Each of these varieties of Hell, real and potential, can be thought of both as an existential challenge and an opportunity to grow or develop beyond that challenge to the next level of challenge. These built-in features of life create The Trouble with Paradise for each of us at an individual level. Much of our personal and collective experience of life is built out of our responses to them.

The following examples of the eight generic kinds of Trouble with Paradise is not meant to be exhaustive—rather it should serve as a jumping off point for you to explore your own versions of these challenges and how you might respond more effectively to them.

1. **Survival—Avoid Predators and other People and Things that can Kill or Injure You**: early on in life we learn to handle primeval fears and often real risks to life and limb. *If we are living in the countryside on a dangerous continent, then there are real natural risks that can be life threatening. In the city, the risks usually involve other people and their bad habits, malevolent ways and much less frequently, the odd serial killer or abductor.* The statistics say you are most likely to be injured or killed in or near a car, and if you are on a bicycle, then you are most likely to be killed by a bus or truck.

    o **Corollary/Response:** Get big and strong, or fast and lean, and learn self-defense, martial arts and play a lot of sport to toughen yourself up. Or buy a fierce dog/pierce your nose/ears if you are a skinny kid whom others kick sand at on the beach.

- **Cultural Exemplars:** Action and Horror Movies, Bodybuilding, Martial Arts, Defense Forces, Guards and Guides of all kinds, Boy Scouts

2. **Security and Belonging—Stay on the Right Side of Powerful and Dangerous People:** through our evolution in tribes, we have learned social skills to ensure we stay on the right side of powerful people, and avoid dangerous people, both of whom could harm us in one way or another. *Sadly in many places it seems that powerful and dangerous people are increasingly one and the same, causing a great deal of damage to others, and most infuriatingly, "get away with it".* Today we practice such skills in our families, communities and organizations to ensure we "go along and get along" or navigate around such people.

    - **Corollary/Response:** Get close to strong or powerful people, or become a strong/powerful person yourself. Manage your alliances carefully. Keep your friends close, and your enemies closer.
    - **Cultural Exemplars:** The Godfather movies, The Sopranos, Wall Street, Traders and most "Investment" Bankers, Superhero comics and movies, tax collectors, local politicians, redneck entrepreneurs, Machiavelli's "The Prince", Thrillers, Good Cops, Bad Cops, Estate Agents and Property Developers, naughty lawyers and accountants.

3. **Finding Balance—Navigating and Reconciling Opposites:** We must navigate and often choose between the opposites that characterize our universe and world, *attempting to find a comfortable balance between them, and suffering when we fail to do so, which is often.* Some find this balance within religious or ideological frames of reference, while others rely on being part of a reliable system that conserves what they know and love (along with quite a lot of what they don't, but they settle for a peaceful life). Others invent their own ways and means of exploring and reconciling opposites through New Age and other contemporary belief systems.

The Trouble with Paradise

- **Corollary/Response:** live a healthy, balanced life avoiding extremes and stick to tried and tested ways of doing things. If you do anything new or venture out into the unknown then check out the risks and take precautions first. Postpone gratification now so as to save up enough to safeguard your future, whether it's your retirement fund or a place in heaven.
- **Cultural Exemplars:** your local savings banker, priests and nuns, managers and bureaucrats, judges, healthcare workers, good lawyers and accountants, "The Firm" with Tom Cruise and other good guys beat bad guys stories where the good guys win by being better at working the system legally, many psychoanalysts, the Catholic Church, Communist Party

4. **Finding Stable Synergy Zones**—We can attempt to find the synergies and symbioses between the different parts of our natural and human worlds through experimentation, innovation and friendly competition, *yet even when we do these synergy zones can be unstable and often decay over time.*

    - **Corollary/Response:** Apply knowledge, skill, science and technology with a dash of humanity to understand and master both natural and human systems to create a better future.
    - **Cultural Exemplars:** systems engineers, scientists, doctors, patent lawyers, the movie "Facebook", ecologists, designers of big systems both natural and human, architects, information technologists, Silicon Valley, great psychologists and therapists.

5. **Designing Thriving Systems:** We can anticipate and design for different possible future worlds, so that we create pleasant and sustainable conditions for ourselves and others, despite the unpredictability of life—*but this takes considerable resources which are not always available to us. Hell is watching others with resources building what you would like to build but can't afford to. Or being forced to remain in a suboptimal environment you would like to change but cannot.*

    - **Corollary/Response:** Create access to or accumulate sufficient resources so as to be capable of adapting to or stabilizing future

life conditions, and ensure that the designs required are suitable for your purposes.
- **Cultural Exemplars:** architects, town and urban planners, integral designers, scenario planners, sustainability experts, sociologists, Masdar City in Abu Dhabi, Transition Towns in the UK, good interior designers like Philippe Starck, Buckminster Fuller.

6. **Mastering the Challenges of Growth and Development:** As we develop and grow through the different stages of our lives, we face more complex challenges at each stage and *there are no guarantees that we will successfully master every challenge; indeed the evidence suggests that we are likely to suffer setbacks particularly at key stages such as adolescence and mid-life.*

    - **Corollary/Response:** Become familiar with developmental models and study the lives of others on a similar journey to yours, so that you can be prepared for the inevitable shifts you will make throughout your life, and enjoy them rather than fight them. Engage in personal development practices that enhance your life and those of others, without becoming a fanatic or dinner party bore about the process.
    - **Cultural Exemplars:** mid-life crises, the movie "About Schmidt", adolescent transition movies, the best personal development methods, Deepak Chopra, a slew of New Age and slightly more old-fashioned gurus, the coaching industry, yoga and all its western derivatives, philosopher Ken Wilber.

7. **Resilience in the Face of Setbacks:** Even when we are living in the best of all possible worlds, *events often supervene which result in shocks, surprises and sudden changes in fortune.*

    - **Corollary/Response:** be prepared by being in good shape and create lots of options that can be activated depending on which setback/shock/change is about to hit you. Learn to find deeper meaning and purpose that transcends day to day concerns and challenges. Find appropriate places in which to recover from

an unfortunate series of events. Make friends when you don't need them.
- **Cultural Exemplars:** the physical fitness industry, spas and retreats, positive psychology, Dr Seligman and authentic happiness, psychotherapy, good insurance policies

8. **Managing our Advancing Years and Final Exit Gracefully:** Finally, however wonderful a life we may have managed to create for ourselves, *we know that it will not last forever and that one day we will die; although our human consciousness finds many ingenious ways to avoid or delay this reality, it is inevitable.*

- **Corollary/Response:** Denial, Staying Alive Longer through Anti-Ageing Methods, Develop Strength of Character to Face the Inevitable, Explore Psychic After Life and Spiritual Phenomena, Engage in projects which leave a legacy beyond your lifetime.
- **Cultural Exemplars:** Empathetic Undertakers, Counselors and Therapists, The Anti-Ageing Industry, Religion, Family, the movie "Amour".

The general Trouble with Paradise from the evolutionary anthropologist/biologist's perspective is that any evolving species is by definition imperfectly adapted to its current niche. That is also the genius of evolution—that we keep on evolving, and that evolution itself evolves.

Take the shark for example: it ceased to evolve 60 million years ago because it became perfectly adapted to its environment. Until very recently, it had no competitors for its ocean scavenging niche. Today, an animal much better adapted to a wide variety of environments is decimating shark numbers: Homo sapiens. We do this in the search of new food sources and shark fins for shark fin soup, while also training sharks to view humans as food through the dubious activity of lowering divers into shark cages trailing pieces of bloody meat for sport and "fun"—the results are all available on YouTube and the Discovery Channel for those of you with a little spare time and a taste for lugubrious adventure.

I was unfortunate enough to experience the consequent adaptation of the shark to such "entertainment" activities while swimming off my favourite beach in Cape Town a few years ago, which resulted in a level one existential lesson. A pair of Great White sharks (the kind the movie "Jaws" made famous), known to be feeding off a small population of seals nearby at Seal Island, had developed a taste for human flesh since shark cages had started to be towed behind boats in the Cape.

As a result the lifeguards, backed up by a helicopter service and shark spotters on nearby mountains, had taken to issuing warnings of shark activity nearby since one or two swimmers and divers had been taken by the Great Whites in the previous few years swimming across the same stretch of water I had swum across since my teens. On this particular day I specifically asked the head lifeguard whether it was safe to swim off the rocks across the bay, and he replied that there had been no shark sightings for a few days, so I should go ahead and dive in.

Still feeling vaguely uneasy, I walked out along the cat walk, stripped down to my swim suit and stepped across the large boulders until I could see the azure blue water below me. I carefully scanned the horizon—everything seemed normal, I adjusted my goggles and took the plunge. The water from the south Atlantic was deliciously cold and refreshing, as I swam crawl toward St James from Sunny Cove.

When I reached the middle of the bay, I stopped briefly to catch my breath and tread water. What I saw next mortified me: it was the most famous scene from "Jaws", playing out directly in front of me, in real life. Everyone was scrambling from the waves toward the beach at high speed. I was gripped by a primeval terror, and then a stream of images flashed before my eyes, recalling everything I had ever read, seen or knew about Great Whites: don't thrash or bleed in the water—it attracts their attention; swim slowly and calmly; if approached face them head on and attempt to punch them on the nose (which is very sensitive); and more.

I began to swim a slow but steady breast stroke toward the beach, praying silently to the universe that I would make it to shore before being swallowed whole by a giant 6m long professional cold-blooded killer. As I made it to the breakers, and swam in to where I could walk, I heaved an almighty sigh of relief: I was alive and still in possession of all four of my limbs!

The next day, an elderly lady was not so lucky. Having been warned by the lifeguards that the Great Whites were patrolling near the beach, she is said to have replied that she had been swimming this route every day for thirty years, and at 70 in the shade she was not going to stop now. After adjusting her bathing cap and goggles one last time, she dived in. She was never seen again. All they found of her was a shredded bathing suit and bathing cap a few days later. Ignoring dangerous aspects of nature can often be fatal, although we all differ in the degree to which we can fully appreciate and work around such dangers successfully.

Pilots have a favourite saying: "There are old pilots, and there are bold pilots. But there are no old, bold pilots". Sharks adapt to changes in their life conditions, and so should we.

Shifting our Responses to Hell on Earth

Sadly for the post-modernists, other ultra-clever intellectuals and New Agers, we only partially construct our reality. I say sadly, because for a narcissist, being able to control reality by being in charge if its construction every moment is the ultimate power kick. And we live in what might be described as the most narcissistic of times, perhaps ever. Just as Narcissus himself stared into that mythic clear pool and drowned after falling in because he became so attached to his own reflection, so too are we attached to our own reflection in the global media mirror that envelops us almost every moment. The danger is not that 21st century mankind will not be connected enough—the much bigger danger is that we will not be able to disconnect.

So, if you too are sufficiently wise not to believe your own online identity (especially your Facebook timeline), then you have learned what all successful public relations advisors tell their clients: whatever you do, do not believe in or fall in love with your own PR. Yet it seems the social media "revolution" is taking our narcissistic tendencies to new extremes. (While we are also developing new capabilities as an emerging global species through our new media, that development is sufficiently complex and critical that we will return to it later).

What does this have to do with evolution and adaptation? Let us start with the dear elderly lady who is no longer with us swimming across the shark infested bay in Cape Town. Perhaps she did not believe that she constructed her own reality entirely, but she was certainly being a tad over confident when declaring that because she had swum across that bay two thousand times before and not been eaten by a shark, that on the 2001'st occasion she could guarantee the outcome would be a success and that she would return to eat lunch in her lovely apartment overlooking the sea.

For her, the shark was an objective reality she simply chose to ignore. Much more sadly for us all, it appears that modern mankind is busy perfecting this tendency to select which aspects of reality it would like to pay attention to, ignoring the warning signs of our age that tell us that if we do not collectively shift our responses to our various hells on earth, that it will genuinely become the equivalent of that sulphur filled inferno which a 21st century Dante would chronicle for posterity.

Perhaps it is no accident that the only other planet in our solar system with any real kind of atmosphere is Venus. The Venusian atmosphere was once similar to ours, but after losing most of the water to space thanks to overheating, today the Venusian atmosphere is comprised primarily of many deadly chemicals such as massive levels of CO^2, nitrogen and clouds of sulphuric acid. The temperature at the surface of Venus averages 467 degrees centigrade and the pressure is 93 bars. To put that in perspective, the air pressure at sea level on earth is a cosy one bar, and the hottest inhabited place on earth, El Aziza in Libya recorded 58 degrees centigrade in 1922, (although 70.7 degrees centigrade was measured in the Lul Desert in Iran recently, but luckily no one lives there otherwise they would have fried a long time ago!)

And the Americans think they are experiencing "extreme weather" when the thermometer hits 107 degrees Fahrenheit or 41.7 degrees Celsius, the current record for July 2012. "Hey wait a minute, isn't that why they call it global warming?", I hope I can hear climate change deniers mumbling quietly to themselves as their electricity bills double to pay for their addiction to air conditioning. A summer holiday on Venus anyone? I thought not.

What would it mean, then, to shift our responses to Hell on Earth? To begin with, we need to take a very long view in the rear-view mirror of our species. The evolution of consciousness was a mixed blessing. It has not been possible to put a precise date on the point at which the first human became self-aware, and shouted "Eureka, that is me in that pond—come look everyone!" Apart from dolphins, elephants some apes and perhaps magpies, most scientists agree that we are a rare example of a self-aware species. It is worth citing Wikipedia's entry on self-awareness theory in full:

"Self-Awareness Theory states that when we focus our attention on ourselves, we evaluate and compare our current behaviour to our internal standards and values. We become self-conscious as objective evaluators of ourselves. However self-awareness is not to be confused with self-consciousness. Various emotional states are intensified by self-awareness. However, some people may seek to increase their self-awareness through these outlets. People are more likely to align their behaviour with their standards when made self-aware. People will be negatively affected if they don't live up to their personal standards.

Various environmental cues and situations induce awareness of the self, such as mirrors, an audience, or being videotaped or recorded. These cues also increase accuracy of personal memory . . . . self-awareness develops systematically from birth through the life span and it is a major factor for the development of general inferential processes. Moreover, a series of recent studies showed that self-awareness about cognitive processes participates in general intelligence on a par with processing efficiency functions, such as working memory, processing speed, and reasoning."

So why would such a wonderful thing such as self-awareness be a mixed blessing? To put it simply, awareness can be both positive and negative. And when we run into one or more of the eight origins of hell we explored at the beginning of this chapter, we can find that the greater self-awareness we possess, the greater pain and suffering we experience. So what can we do if we are experiencing the kind of great pain and suffering that cannot be turned off? There are several options, from zapping the source of pain and suffering itself, drugs, meditation and getting the hell out of wherever you are.

One favourite, tried and tested solution of our species is to get a dose of strong religion. This built-in tendency of our species reveals a great deal about us, our ability to believe almost anything we want to believe, and the power of our minds over matter.

Dr Robin Lincoln Wood

A Short History of Strong Religion and Evolution

Strong religion appears to have been a natural feature of Homo sapiens for tens of thousands of years. Our growing brain along with its growing awareness demanded an explanation for the seemingly uncontrollable events around us, and the faeries, trolls and gods that conveniently took human form and took a handy interest in our affairs, fitted the bill nicely. Initially it seems that we spent a great deal of time and effort attempting to appease the gods that could cause us harm (such as Thor, the god of thunder), along with entreating the gods that could bring us good to do so, such as the gods that brought the annual rains for our crops.

Magical thinking is a type of causal reasoning that looks for meaningful relationships between acts and events. It enables us to enjoy Walt Disney, Lord of the Rings and Harry Potter even as adults. This sense of awe and wonder is vital to the development of our minds and also gives us hope in the future. Yes, I can be powerful! Yes, I can control things and get what I want! Yes, there is hope for me!

In religion, folk religion, and superstition, believers assume a strong causal link between religious rituals, (such as prayer, sacrifice, or the observance of a taboo), and an expected benefit. In clinical psychology, magical thinking is a condition that causes the patient to experience irrational fear of performing certain acts or having certain thoughts because they assume a correlation between their actions and impending calamities. Obsessive compulsive disorder is also closely linked to magical thinking, in that OCD sufferers reduce their anxiety levels through the ritual repetition of a particular sequence of behaviours—for example washing their hands with a special bar of soap over and over, as Jack Nicholson so beautifully portrayed in "As Good as it Gets".

Sometimes our efforts to appease the gods can themselves be quite frightening. In Africa today, there are still "witches" being burned or killed when lightning strikes a village or a hut. The "witch" usually turns out to be a toothless old harridan who has incurred the wrath of some of the villagers, and is therefore ripe for the picking. Witch doctors known as sangomas are still highly popular in treating all kinds of ailments, and sometimes there is a degree of efficacy in their remedies when based on a deep knowledge of locals' plants and their healing effects. And who

The Trouble with Paradise

can forget the Mayans with their live human sacrifices to the gods to ensure a bountiful harvest?

In short, our brains are naturally wired for magical thinking, and this is a normal stage of development in Homo sapiens from between the ages of five and twelve. Magical thinking is what we do when we are children. It is the type of thinking that produces the memorable moments in your life such as of little kids playing hide and seek. If the child covers her eyes then no one can see her—even if she is standing in the open. It is also the type of thinking that produces the Tooth Fairy and Santa Claus. Before you send out the lynch mob let me be very clear. Up until age 8-10 Santa is real—I certainly believed so and I certainly believed that my kids believed it too.

For this age group magical thinking is perfectly fine, normal and natural. In fact, it is necessary in order for them to develop a concept of self and other; it their way of learning about the world. At some stage we grow up, and our beliefs change. Our world expands and, sooner or later, we realise that the map is not the territory.

The trouble with magical thinking is when it persists past the age of 12. If you buy into it you cannot win. Why? Well most obviously it does not work and you will find out soon enough if you are open to questioning your outcomes. At its worst, it carries a very nasty sting in the tail which will have you believing that it is your fault if your dreams do not come true, because you did not have the "faith" in its power. The classic double bind—you can't win. It also makes it very challenging to point this out to the subscriber to the magical thinking model. They simply point out that you are a cynical unbeliever.

The Fundamentalism Project is a global study begun after the fall of the Twin Towers on 9/11/01. It is a decade-long interdisciplinary study of antimodernist, antisecular militant religious movements on five continents and within seven world religious traditions. The authors of this study analyse the social structures, cultural contexts, and political environments in which fundamentalist movements have emerged around the world, from the Islamic Hamas and Hezbollah to the Catholic and Protestant paramilitaries of Northern Ireland, and from the Moral Majority and Christian Coalition of the United States to the Sikh radicals and Hindu nationalists of India.

Strong religion apparently comes in several different modes of fundamentalism, and is usually triggered by dramatic historical events from earthquakes to famines to wars to economic crises. It can also run in families as a tendency, as I know it does on both sides of my family. (Luckily we seem to be having an intergenerational break from strong religion in our particular branch). There are many underlying similarities between all fundamentalist groups, though not all of them represent a potential threat to modern, secular society and religious pluralism and tolerance as do the jihadists and extreme right wingers in modern democracies. Many are good people who go to church regularly, believe the world was created in seven days around six thousand years ago, and help the sick, old and poor.

Globally most Christians believe that the human race and nature emerged through evolution (including the highly influential Pope), but in countries such as the USA, most Christians are also creationists. Islamic, Hindu and other faiths also have their own versions of creationism.

Given the global scientific consensus around evolution, why would anyone choose to believe in a fundamentalist religion that flies in the face of two centuries of hard evidence, you might ask?

Two centuries ago, almost the entire population of Great Britain believed in creationism—there was no credible alternative that had any mainstream support. Then, along came Charles Darwin. Although Darwin's father was a medical doctor, Darwin's own studies at Edinburgh Medical School were unsuccessful due to his distaste for surgery and lack of interest in many of the lectures. In his second year of his medical studies, Darwin was exposed to the evolutionary ideas of Lamarck, whose ideas were remarkably similar to Darwin's grandfather Erasmus. Darwin handily also learned some geology and how to classify plants.

The neglect of his medical studies led Darwin's father to send him to Cambridge for a BA so that he could become a parson. Instead young Charles became a friend of botany Professor Henslow and began to develop a talent for natural philosophy and history. Henslow secured a place for Darwin on the HMS Beagle to map the coastline of South America. The rest, as they say, is history.

After the Beagle's return to England in 1836, Darwin rapidly became a celebrity scientist. Thanks to funding from his wealthy family, he was able to deploy a team of scientists to catalogue and classify all of the specimens he brought back from his five year voyage. Darwin later noted in his autobiography:

The Trouble with Paradise

"In October 1838, that is, fifteen months after I had begun my systematic enquiry, I happened to read for amusement Malthus on Population, and being well prepared to appreciate the struggle for existence which everywhere goes on from long-continued observation of the habits of animals and plants, it at once struck me that under these circumstances favorable variations would tend to be preserved, and unfavorable ones to be destroyed. The result of this would be the formation of new species. Here, then, I had at last got a theory by which to work."

By 1859, after many personal setbacks and the death of his young son from scarlet fever, Darwin's book "On the Origin of the Species" was finally published and sold out before it even hit the bookshelves. Although Darwin had avoided using the word "evolution", and refrained from making any connection between the origins of man and his new theory backed up with hard evidence, the Church of England split over this new idea. It was to be the beginning of a centuries long debate which, though Darwin's work has been proven beyond a shadow of a doubt by 150 years of hard scientific evidence, the idea of evolution itself still ignites controversy amongst religious fundamentalists.

It took Darwin another 12 years to research and write "The Descent of Man", in which he concluded from all the available evidence that while human beings are all one species, we are in fact, animals. In his last book on this topic: "The Expression of Emotions in Man and Animals", Darwin's conclusion was:

"that man with all his noble qualities, with sympathy which feels for the most debased, with benevolence which extends not only to other men but to the humblest living creature, with his god-like intellect which has penetrated into the movements and constitution of the solar system—with all these exalted powers—Man still bears in his bodily frame the indelible stamp of his lowly origin."

By the time of his death from heart failure in 1882, Charles Darwin was a national celebrity and internationally renowned. In recognition of Darwin's pre-eminence as a scientist, he was honoured with a state funeral and buried in Westminster Abbey, close to John Herschel and Isaac Newton. He is today remembered as one of the most influential figures in human history, Today, 80% of the population of Western Europe believe evolution to be true, which is several hundred million people more than did so two centuries ago, thanks to the work of Charles Darwin.

In the USA, however, recent research by the National Geographic Society found that substantial numbers of American adults are confused about some core ideas related to 20th—and 21st-century biology. The researchers cite a 2005 study finding that 78 percent of adults agreed that plants and animals had evolved from other organisms. In the same study, 62 percent also believed that God created humans without any evolutionary development. Fewer than half of American adults can provide a minimal definition of DNA, the authors add.

Might it be time for a decent education system in the USA? Spending six times more on prisons than on education and a trillion dollars a year on military and related spending which exceeds the entire budget for education at all levels in the USA might seem a strange set of priorities for a country which loves to tell its youngsters that it is the "greatest country in the world".

Internationally, it appears that ignorance is the main cause of a belief in creationism. A recent survey of 10 000 adults in ten major world countries found that knowing meant believing in evolution. Fifty-six percent of the people in all 10 countries who had heard of Darwin believed there is sufficient scientific evidence in support of Darwin's theory of evolution. A more detailed analysis, however, revealed a complex picture. Although the majority of adults surveyed in India (77%), China (72%), Mexico (65%), the United Kingdom (62%), Spain (61%), and Argentina (57%) accepted the theory of evolution as scientifically founded, only 48% did so in Russia, 42% in South Africa, 41% in the United States, and 25% in Egypt.

So, overall, it appears that if people are actually exposed to a scientific education and general knowledge, they then tend to go with the evidence, which is heartening, despite strong religion still having about half of the human population firmly in its grip. Given that our species is still going through its adolescence, as it grows into adulthood we can hopefully expect it to get a better grasp of the vast body of knowledge available and to use that for the advancement of all humankind. Though don't hold your breath, it might take a few more centuries. The only question is: will we make it through that period as a species? Is it already too late?

We will return to re-examine Darwin's assumptions and discover that although he was right in general about evolution, his dark assumption about scarcity being the only driver of evolution was only a small part of the picture.

The Uses and Abuses of Self-Awareness

As we saw earlier, being self-aware can be a mixed blessing. In fact, being hyper self-aware and highly sensitive can be its very own form of hell. If one of the most natural responses to anxiety, fear, uncertainty and doubt has historically been to get a dose of strong religion, then it takes a brave person indeed to step out of the shadows of this sometimes beneficial but very often corrosive tendency toward magical thinking.

There is a spectrum of self-awareness that begins with a basic awareness of one's own body and emotions. For example, some people struggle to regulate their temper and "fly off the handle" easily, while others appear to stay calm and composed. In the case of anger management, some of the challenge stems from innate differences between people genetically—recent research reveals that the hotheads are cognitively impaired in their ability to mentalize the emotions both of themselves and others. Not only do they experience much higher distress levels than cool cucumbers, but they can also struggle to take the perspective of others.

Luckily, self-awareness and empathy can be taught, so that apart from the most challenged individuals at the extreme hothead end of the spectrum, most low self-awareness challenged people can learn to get in touch with their own emotions and those of others over time.

In the middle lane of self-awareness, we find people who are natural socializers and empathizers. They have become highly skilled at balancing the extremes of emotions and make good negotiators and dispute resolution facilitators. In the fast lane of self-awareness are those people known as "highly sensitive", who process sensory data much more deeply and thoroughly due to a biological difference in their nervous systems. "HSP's", as they are known, comprise about a fifth of the population, and can become shy or inhibited depending on the environment.

The opposite of an HSP is an HSS: a "High Sensation Seeker", who takes many risks, and acts without reflecting deeply or at all. Interestingly, an HSP can also be an HSS, but such a hybrid won't take any unreflected-upon-risks. Apart from being more sensitive than the rest of us HSP's may also suffer inwardly from over stimulation from levels of sensory input that others find completely acceptable. HSP's are therefore more sensitive to pain, easily overwhelmed by sensory input and aesthetic sensitives are deeply upset by ugliness and unnatural environments.

One of the defenses against being overwhelmed by sensory input is to create or be drawn to orderly environments where everything can be neatly arranged. Although sensitives share a tendency to be easily over stimulated with children on the autistic spectrum, these conditions are different as under the right conditions sensitives can display excellent social skills, while someone on the autistic spectrum cannot.

Having the largest brain in the animal kingdom proportional to our weight (about 3% on average for a human, compared with 0,034% for the large whales whose extremely large brains are a much smaller percentage of their body mass) means human beings experience anxiety and depression on a massive scale compared with the rest of our animal brethren. Anxiety and depression are two of the most common forms of mental distress, and can be triggered by any number of events and causes—the birth of a child, long dark winters, stress at work or in a relationship, moving to a new home in an unfamiliar environment, and so on.

Some anxious people worry themselves and others to death, while other have frequent panic attacks or harm themselves. Many have trouble sleeping, or do very badly at exams due to their nerves. While antidepressants, sleeping pills and tranquillizers can help in the short term, behavioural intervention or cognitive therapy is required if the sufferer is to make a sustainable recovery.

In the past few decades psychology has gone from treating mental illness and suffering, to focusing on how we can raise the bar of the human condition. Two popular words in today's psychology lexicon are "thrive" and "flourish".

The blurb for Dr Martin Seligman's latest book, "Flourish", puts it like this:

"What is it that enables you to cultivate your talents, to build deep, lasting relationships with others, to feel pleasure, and to contribute meaningfully to the world? In a word, what is it that allows you to flourish? "Well-being" takes the stage front and center, and Happiness (or Positive Emotion) is one of the five pillars of Positive Psychology, along with Engagement, Relationships, Meaning, and Accomplishment—or PERMA, the permanent building blocks for a life of profound fulfillment."

This sounds like, and probably is a noble project that will make the world a better place, but let us put our black hat on for just one moment. While flourishing is clearly worthwhile goal in its own right, might there be some downsides to some of its components? For example, we are told that being positive is the key to success and prosperity.

Yet must being positive come at the expense of truth and facing up to reality? If we have a problem or face a challenge, isn't it better to be realistic about the actual situation we are in and the nature of the problems we need to tackle? Was the recent global economic crash, triggered by very positive, but also fraudulent Wall Street traders, not a warning?

False promises of positive thinking reach into every corner of modern life, from Evangelical mega churches to the self-improvement industry to the medical establishment, and, worst of all, to the business community, where the refusal to consider negative outcomes—like mortgage defaults—contributed directly to the recent global economic depression. The downside of positive thinking in this case are clear—the average man in the street was robbed, the wealthy players who perpetrated the con got wealthier, and: the guilty got away scot free while the innocent were punished. By being relentlessly upbeat (to the point of becoming self-delusional), is it possible that we are missing out on what is authentic and real? Are we missing risks we should actually be worried about and acting to prevent?

In a Nazi concentration camp, during the darkest days of the Second World War, an Austrian neurologist and psychiatrist deported to the Nazi Theresienstadt Ghetto was assigned to head a mental health care unit for the ghetto. Though he lost his father there, and then lost his wife later on at Bergen Belsen while he was transported to Auschwitz, Frankl stayed alive by helping others survive the atrocities of this man made hell on earth. Over the years, he noticed that those who were most likely to live, were those who had something to live for, a purpose or meaning in life. In 1945 he wrote his famous book: "Man's Search for Meaning", in which he gives an example of his observations in an account of his own experience in the harsh conditions of the Nazi labor camps:

"We stumbled on in the darkness, over big stones and through large puddles, along the one road leading from the camp. The accompanying guards kept shouting at us and driving us with the butts of their rifles. Anyone with very sore feet supported himself on his neighbor's arm. Hardly a word was spoken; the icy wind did not encourage talk. Hiding his mouth behind his upturned collar, the man marching next to me whispered suddenly: 'If our wives could see us now! I do hope they are better off in their camps and don't know what is happening to us.'

That brought thoughts of my own wife to mind. And as we stumbled on for miles, slipping on icy spots, supporting each other time and again, dragging one another up and onward, nothing was said, but we both knew: each of us was thinking of his wife. Occasionally I looked at the sky, where the stars were fading and the pink light of the morning was beginning to spread behind a dark bank of clouds. But my mind clung to my wife's image, imagining it with an uncanny acuteness. I heard her answering me, saw her smile, her frank and encouraging look. Real or not, her look was then more luminous than the sun which was beginning to rise.

A thought transfixed me: for the first time in my life I saw the truth as it is set into song by so many poets, proclaimed as the final wisdom by so many thinkers. The truth—that love is the ultimate and the highest goal to which man can aspire. Then I grasped the meaning of the greatest secret that human poetry and human thought and belief have to impart: The salvation of man is through love and in love. I understood how a man who has nothing left in this world still may know bliss, be it only for a brief moment, in the contemplation of his beloved. In a position of utter desolation, when man cannot express himself in positive action, when his only achievement may consist in enduring his sufferings in the right way—an honorable way—in such a position man can, through loving contemplation of the image he carries of his beloved, achieve fulfillment. For the first time in my life I was able to understand the meaning of the words, "The angels are lost in perpetual contemplation of an infinite glory".

Frankl concluded that the lack of meaning is the paramount existential stress. To him, existential neurosis is synonymous with a crisis of meaninglessness. He is thought to have coined the term "Sunday Neurosis" referring to a form of anxiety resulting from an awareness in some people of the emptiness of their lives once the working week is over. His diagnosis of complaints of a void and a vague discontent was that this arises from an existential vacuum, or feeling of meaninglessness, which is a common phenomenon and is characterized by the subjective state of boredom, apathy, and emptiness. One feels cynical, lacks direction and questions the point of most of life's activities. Sound familiar to any of you out there?

Frankl once recommended that the Statue of Liberty on the East Coast of the United States be complemented by a Statue of Responsibility on the West Coast:

"Freedom, however, is not the last word. Freedom is only part of the story and half of the truth. Freedom is but the negative aspect of the whole phenomenon whose positive aspect is responsibleness. In fact, freedom is in danger of degenerating into mere arbitrariness unless it is lived in terms of responsibleness. That is why I recommend that the Statue of Liberty on the East Coast be supplemented by a Statue of Responsibility on the West Coast."

There are reportedly plans to construct such a statue. Perhaps now would be a particularly meaningful moment for all of us?

A "Good Enough" Modern Paradise?

There is no doubt that life is, can be, and has always been demanding; tough; and, occasionally murderously challenging. Along with our self-awareness, technologies and culture, human beings have had to evolve coping mechanisms to handle both predictable troubles in paradise, as well as unpredictable shocks and surprises. Both religion and science have evolved directly in service of the fundamental human need for certainty, order, hope and their beneficial role in our survival.

Psychologist Abraham Maslow's hierarchy of needs predicts that we cannot focus on our higher needs if we are lacking fundamentals such as security, shelter, food and belonging. I have had a few "spiritual" people contradict me on this: they apparently believe that one can be spiritual even if starving in a gutter somewhere. I am absolutely certain these "spiritual" people have absolutely no personal experience whereof they speak. They also confuse what might be a temporary spiritual state of a saint or martyr on a fast or burning in the flames, with the much longer term phenomenon of human development.

As we discovered earlier, there is an undeniable trajectory in nature toward greater complexity, symbiosis, consciousness and collaboration. The emergence of each of these phenomena has occurred so many times in parallel across so many different species and locations for so many billions of years as to be irrefutable. This developmental trajectory can be seen in the human being from the moment of its conception as a fish-like foetus, to its development of reptilian and amphibian features, and then finally the mammalian features emerging fully along with our uniquely large brain in the last few months of pregnancy.[3]

The mammalian brain is what makes it possible to make the leap from fish-like passivity through reptilian/amphibian dominance and aggression to mammalian curiosity, playfulness and learning. Every day we each make that transition from deep dreamless sleep and the trance like state of a fish, to the semi-reptilian state of dreaming and waking up, to the fully conscious state required for us to do our work and live our lives to the full.

Our large brains, opposable thumbs and the ability to walk, run, hunt, gather, cultivate and build have enabled us to become the dominant species on earth in a remarkably short evolutionary time span. All of this happened in the material world, and was accelerated by the scientific and technological revolution catalyzed by the first Renaissance from 1500 onward in Europe. Unlike earlier breakthroughs in China, due to the open nature of the European system this revolution has now spread around the entire planet.

According to many different calculations, Homo sapiens reached an "abundance point" in the developed world during the 1950's thanks to our exploitation of technology and natural resources. The standard of

---

[3] As the biologist Gregory Bateson put it, Ontogeny recapitulates phylogeny— or put more simply, the very structure of every animal repeats the sequence of every stage of development of that animal. The more advanced the animal, the more stages there are.

living attainable by an unskilled factory worker at that time exceeded what would only have been possible in prior ages through colonialism and slavery. The fossil-fuel powered machines now did the work slaves used to do, from supplying water to carrying and burning fuel to cultivating and harvesting crops to building highways and cities to making consumer goods. We had never had it so good, our politicians said, and it was true.

The new levels of material comfort and security enjoyed by all classes and stations of life over the past sixty years have been historically unprecedented, to the extent that the baby boomer generation born after the end of the Second World War in 1945 grew up in a material paradise. They were the first generation never to go hungry, or lack the basics such as food, shelter, clothing and community amenities. They also turned into the world's first "teenagers".

There were no "teenagers" before World War II. Instead of Teenagers, there were Youths. Youths were young people who wanted to become adults. However confused, wayward, or silly they acted, however many mistakes they made, they looked to the future. They knew that adult life was different than a child's life. They planned to grow up, leave childhood behind, and become adults. They were aware that life is more than youth

Previously human beings between childhood and adulthood were called kids, boys and girls, young people, adolescents, and youths. These young human beings were addressed as "Young man" and "Young woman." Looking at them, their parents thought, "My growing son," and "My growing daughter," and they addressed them as "Daughter" and "Son." Sometimes others addressed them as "Master" and "Miss." Even the words "gentleman" and "lady" were sometimes heard. The word "teenager" did not exist. Compare the entries in Webster's Second (1934) and Third (1961) editions; only after the war does the adjective "teen-age" become the noun, "teenager."

For the very first time, for the average citizen rather than a tiny elite or royal families and nobility, this age of abundance produced a generation that could, and did, take the basics for granted. And as an outlet for their abundant youthful "teenage" energy, accelerated by the emerging music, film, television and advertising industries, we saw the birth of rock 'n roll, fast food, discos, political activism and the arrival of an Eastern idea whose time had come.: "Discovering oneself", "experiencing enlightenment" and "tune in, turn on and dropout" had arrived.

At the forefront of this new movement were a young group of musicians from Liverpool. Their rhythm and lead guitarists jumped in with both feet, so to speak, and started a trend which continues to grow to this day. Their names were George and John.

The Beatles first met the Maharishi Mahesh Yogi in London in August 1967 and then attended a seminar in Bangor, Wales. In February 1968 they visited Rishikesh in India to attend an advanced Transcendental Meditation (TM) training session at the Maharishi's ashram. Their stay at the ashram was one of the band's most productive periods and received widespread media attention. Some credit their adoption of the Maharishi as their guru as changing attitudes in the West about Indian spirituality, and encouraging the study of Transcendental Meditation.

A 2007 American government study found that nearly 10% of the American population, 20 million people, had meditated in the previous twelve months. Similar figures would be expected in Western Europe, where meditation was introduced at much the same time as the USA in the 1960's.

Transcendental Meditation (TM) became one of the most popular meditation methods in the west, although the Maharishi developed a reputation as a controversial character due to his accumulation of very large amounts of money and his many affairs with female students. He once tried to persuade the Beatles to give him 25% of the proceeds from their next album, but John Lennon was reputed to have replied, characteristically frankly: "Over my dead body". Although George Harrison and his wife continued to maintain a connection with Maharishi, Lennon made a very public denunciation of his shortcomings. George Harrison became a particularly devoted daily meditator, and even built a special meditation room in his English manor house at Friar Park.

There are so many different varieties of meditation that, as with sport, one would have to classify them into many categories. The one thing they all have in common, though, is their focus on the breath and on emptying the mind of thoughts to focus on emptiness through a variety of techniques. What one discovers upon entering into this emptiness, is an interior world which is seldom examined in our western, materialistically focused culture.

Meditation is now catching on as a very effective method of relieving stress and improving health and cognitive functioning. Thousands of studies over the past century have shown that meditation reduces the incidence of disease, injuries and even crime where group meditation is practiced. These studies of people who meditate report many real benefits:

- Heart rate, respiration, blood pressure and oxygen consumption are all decreased.
- Meditators are less anxious and nervous.
- Meditators were more independent and self-confident
- People who deliberated daily were less fearful of death.
- 75% of insomniacs who started a daily meditation program were able to fall asleep within 20 minutes of going to bed.
- Production of the stress hormone Cortisol is greatly decreased, thus making it possible for those people to deal with stress better when it occurs.
- Women with PMS showed symptom improvements after 5 months of steady daily rumination and reflection.
- Thickness of the artery walls decreased which effectively lowers the risk of heart attack or stroke by 8% to 15%.
- Relaxation therapy was helpful in chronic pain patients.
- 60% of anxiety prone people showed marked improvements in anxiety levels after 6-9 months.

It has also been documented that people who use meditation and relaxation techniques may be physiologically younger by 12 to 15 years—quite amazing results given that after several decades of jogging and aerobics fans of these sports have discovered that they may only increase life span by an average of six years.

Yes, meditation, along with medication and religion, can reduce suffering and stress, help you live longer and produce many other benefits—but can it make you happy and satisfied with life? And does it produce a better world? On its own, probably not, though in combination with other approaches and practices it could be an important ingredient of a fulfilling life.

If meditation, mindfulness and reflective practices are an excellent way to begin reducing stress and pain, where do we go from there? What would your own corner of paradise look like? And how would you get from here to there? Let's get into our psychological helicopter and see the world as a system of interacting perspectives that enables us to harness and develop our four core intelligences: our bodies (kinaesthetic), emotions (emotional), minds (cognitive) and spirits (spiritual).

The World as a System of Interacting Perspectives: From "I", to "IT" to "WE" to "ITS"

Philosopher Ken Wilber begins his system of integral philosophy with the "I" that we become aware of in this meditative state. Western philosophy has largely been focused on the exterior, material world since the time of Aristotle, while Eastern philosophers have focused more on our interior experience of ourselves and our relationship to the exterior world. In fact it was not until the emergence of psychology during the 19th century that westerners could legitimately apply a framework of any kind to their inner experiences beyond those supplied by religion such as prayer and reflection.

This first person "I" forms the inside of our interior experience, and is also what cognitive scientists and materialists recognize as the source of "qualia", or our raw subjective conscious experience. This is our "**Personal Interior**".

The famous quantum physicist Erwin Schrödinger had this counter-materialist take: "The sensation of colour cannot be accounted for by the physicist's objective picture of light-waves. Could the physiologist account for it, if he had fuller knowledge than he has of the processes in the retina and the nervous processes set up by them in the optical nerve bundles and in the brain? I do not think so."

"Redness" is an example of a "quale" (singular of qualia). We can both recognize redness when see it, although to a colour blind person redness does not exist. We could get tied in a tangle of knots discussing the kind of experience we all take for granted every waking moment, so I will resist the temptation to dive into the extensive debates that have raged between philosophers, cognitive scientists and scientists around qualia. For our purposes, what matters is that both you and I know that we have our own subjective experience, which is quite often

The Trouble with Paradise

very different from the subjective experience of others, even if we are experiencing the same phenomenon. For example, I might like a big red wine with a strong finish, while you will find the same Cabernet Sauvignon very heavy and less to your liking than, say, a smoother (but to me very bland) Merlot.

Now, there may never be any objective way for you and I to compare our exact taste sensations of the wines at our wine tasting, but neither of us can deny that the other has a different inner experience of the wines. Some thinkers believe that our mental experience and physical reality are two distinct, separate aspects of reality, and Wilber's thinking starts with a variation of these two aspects of reality arising together, or co-arising, as a single whole, or "Holon"[4].

It would make sense in a universe where everything is interconnected, that our interior experience and our external world arise together. There is a direct cause and effect relationship between my intentions and motivations and what actually happens. Sometimes this relationship is less direct and more ineffective than we would like it to be, especially when we operate from a distorted or over-simplified view of reality. Equally, when I stub my toe in the material world I immediately have an experience of pain which I express as "Ouch!" We can all relate to that, without too much philosophizing.

Whatever I discover inside the interior of my own consciousness, however, I automatically interpret that experience in terms of my own worldview, which is formed from all of the experiences and learning I have ever had. Your and my worldview will inevitably be different as we are separate individuals leading different lives with different backgrounds. Even identical twins develop slightly different personalities and worldviews, though they are brought up in exactly the same surroundings.

---

[4] After a lifetime having this debate with myself and others, I have come down against the radical materialists such as Dan Dennett who believe that because there is no way to fit subjective experience into modern scientific theory, we must reject qualia as unscientific. That is simply a very poor/non-argument., and every one of Dennett's thought experiments can be countered with an equally powerful counter example. And at that point the entire debate becomes absurd and rather comical.

This is the "outside" of your personal interior, which translates the experiences you have on the "inside" of your interior, into thoughts, conclusions, responses, actions and the "real world".

The objective, second person "IT", in contrast with the first person, subjective "I", resides in our bodies and brains, much beloved by the medical profession and cognitive scientists, because they can be studied objectively, and, for the materialists, are therefore more real than our inner experiences. This is our "**Personal Exterior**".

How frustrating it must be for these "hard" scientists to be unable to simplify everything in the universe down to organs, cells, DNA, molecules, atoms and quarks. Scientists and thinkers who insist that everything can ultimately be understood in these terms by being broken down into its parts, are known as reductionist materialists. They are often a sad lot, given that life has no real meaning or purpose for the vast majority of dedicated materialists, so their experience of life can often end up being rather superficial.

Your and my personal exteriors are obvious and highly visible if we are in the same room together, but our personal interiors are hidden, and being able to make them visible would depend on the extent that we are both consciously aware of their contents and able to share those thoughts or emotions adequately with each other.

Our personal exteriors comprise brains, hearts, guts, muscles and a vast number of other body parts organized into several major organ systems built out of between 210 to 411 cell types, (depending on how they are classified), neatly packaged into a human body capable of some amazing feats. Most of the evolutionary sciences have focused their efforts on studying bodies old and new, simply because there is so much evidence to show how our bodies have evolved in response to changing life conditions. In order to understand how our minds, consciousness and culture have evolved, however, we have to study the artifacts we have produced over the millennia from cave paintings to vast urban civilizations.

But no man is an island—we are born into families and communities, and develop and grow in a variety of social institutions from schools to universities before embarking on our careers and adult lives. Being social animals, we are "socialized", and our learning and experiences build on everything that we take for granted in our youth. Which brings us to the "WE".

On our own you and I are "I"'s and "IT"'s, but if we connect with each other in some way together we also form a "WE". Bumping into each other in a crowd by accident would not create a "WE", but as soon as we recognize and acknowledge each other's presence, we have formed the very first layer of a relationship. It may all amount to nothing, and we might never see each other again, but a connection has been made.

If you and I have grown up in the same culture, speaking the same language, then we will share a great deal of background, even if we are complete opposites in every way. This background enables us to not only communicate, but also to share meaning, understand the significance of different roles and tendencies of people in our shared world, and even predict some of the things people will do and say—all thanks to a shared cultural background. This kind of WE is also referred to as our **Collective Interior**.

While the felt connection of the WE is the inside of our Collective Interior, the syntactical structures and basic building blocks of our culture operate as the outside of our culture, shaping what it is possible to articulate and communicate through our culture. These structures are deeply embedded in us and largely taken for granted until they become the object of study or emerge during a breakdown in communication between us—when one of our WE says "WHAT?!"

If, however, we are from alien cultures and do not share a common language, we will have to rely on common reference points to communicate in a kind of sign language, which can vary from being amusing to downright dangerous in certain situations if significant misunderstandings arise. Not only will a "WE" never arise, but if things do not go well we might swiftly recruit others to create "US" and "THEM" on both sides, a well-known starting point for conflict generation in human history.

At the other extreme, there are many well-intentioned people that love to declare that "WE are all ONE". If the WE referred to is a group of people around the world who share similar beliefs, for example, Christians, and they all spoke English, then that would be a good starting point for their "ONE". Given that there are over 6 000 languages spoken around the world, located in a wide variety of cultures, this assumption of WE are ONE ranges from highly ambitious to very naïve.

Yet, at some level, I can appreciate what the WE are ONE folks are trying to say: that we human beings, as a species, are remarkably similar in many ways, sharing a dozen basic emotions, facial expressions and basic values, and that arguing about our differences is not terribly productive. My own view, however, is that by recognizing both differences and similarities, we are much likelier to end up with a much greater range of mutual understanding and common action than if we simply assume WE are all ONE from the get go without taking into account our diversity and uniqueness.

Finally, we arrive at our **Collective Exterior**, or "ITS", where we co-create the world around us using technologies, both "hard" and "soft". Our collective exterior is made up of all the infrastructure and hard systems we have created over millennia, from cities and transport systems to educational, healthcare, defence, business and communications systems. This collective infrastructure is basically what would be left after a neutron bomb was dropped—only the people would be gone. Such a comforting thought.

The astonishing thing is that it is this collective exterior that we use to measure our "progress" as a civilization, by and large. We call the things that make up this collective exterior "assets", and what we pay to the people who build and maintain this material world are called "expenses", while the biggest lie ever told by CEO's is that "people are our greatest asset". That lie has worn so thin that CEO's do not even bother to roll it out for the press any more.

It is one short step from the idiocy of this industrial age accounting logic, to an economy where workers become wage slaves with declining real incomes while top management become multi-millionaires and investors billionaires because the latter are good at "sweating the assets" and "cutting costs" (almost always the payroll). This has turned into a global negative sum game where we have now, thankfully, run out of Chinese prison workers and Indian child labourers who can take jobs away from workers in the developed countries.

In turn, this reduces the amount of money available for consumer spending in developed economies, which then reduces the turnover of businesses in developed countries, so that their only option is to sell more goods and services abroad in developing nations. And we wonder why the economies of the G8 have stalled, permanently it seems.

Of course, this collective exterior, our modern global infrastructure, our frenetic cities crammed full of shopaholics, our suburbs full of strip malls, hypermarkets and car dealers, is built on an unsustainable model of permanent economic growth. In other words, everyone has to keep buying more and better stuff, even when their houses are so full of stuff they have to rent storage elsewhere to accommodate the overflow. (One of the world's fastest growing markets is in storage for homeowners and businesses).

We may have completely lost the plot. In many developing countries, for example, 90% of work and economic output is not measured, simply because it is done for love and not money. Parenting, home healthcare and education, childcare, nursing of the elderly and infirm, home cooking, backyard vegetable gardens, building one's own home, and much more, are simply outside of the official system. What Hazel Henderson calls the "love economy" is alive and well and keeps people alive and well in well over 150 countries around the world.

Although I am not going to suggest that life for people in the informal economy in developing countries is a bed of roses, there is a relative absence of our modern diseases of obesity, anxiety, stress-related illnesses, drug addiction, crime waves, urban pollution and endless traffic jams. In some cases, communities in these places also exhibit greater longevity than those paying a fortune for healthcare in the modern world.

I am also not suggesting that we can all go back to live in some kind of mythical pre-modern Eden. That would neither be realistic or truthful. There are many dangers in the pre-modern world, from infant mortality to deadly diseases to natural hazards, and often a risk of serious poverty and hunger. There are threats from local warlords, witch doctors, property developers, corrupt politicians and police, loggers and miners. But there are enough places where a fusion of pre-modern sensibilities and modern technologies are well blended and conducive to wellbeing, to demonstrate my point.

So, while the Paradise beloved of modernists who believe that all our problems can be solved with technology is just a pipe dream, we find that the negative consequences of development by people who believe this pipe dream, are very real.

While out of control development is both highly damaging and carries huge long-term costs, it is also completely unnecessary in the absence of greed and ignorance. There are so many wonderful examples today of well thought through, sustainable developments in country after country, that we need simply look, listen and learn from what is already working well.

We are still globally in the grip of a small clique of very rich and very powerful people who control 80% of the assets of the world, as well as buying control of legislatures and governments everywhere. This problem has reached epidemic proportions in the USA for example, where Congress is essentially a corporate mouthpiece arguing for anything that can produce greater corporate profits. Everyone knows the system is broken, but the powerful and greedy are not going to give up their wealth and power without a fight, despite what Occupy Wall Street and other protest movements wish.

In 2000 the UN calculated that the richest 10% of adults in the world own 85% of the world total, whereas over a decade later the super-rich in the USA own more of the national wealth than at any time since 1928. The richest 1% in America own one third of all the wealth in the USA, while one in seven Americans lives below the poverty line. That is 46.2 million people in the world's "richest" country who do not have a regular supply of food.

Of course, ever since the economist Vilifredo Pareto coined the 80/20 law in 1906, we have been well aware that 80% of the wealth in any natural system is owned by 20% of the system. Pareto originally calculated that 80% of the land in Italy was owned by 20% of the people, and 80% of the peas in his garden were contained in 20% of the pea pods. In business it is also largely true that 80% of sales come from 20% of the customers in most businesses.

Physicists call this the "Power Law", because it also describes an 80/20'ish relationship between the frequency of an event and its size. For example, the number of cities with a certain population size, the size of earthquakes, solar flares, neuronal populations, words, species riches, power outages—all follow a power law that says that 20% of events or instances account for 80% of the occurrences above a certain size. Power law population relationships are may be a result of the structural dynamics of growth. In other words, "winners-win" patterns emerge (the rich get richer until some sort of catastrophic failure or limit to growth is reached).

There is, of course, much evidence that suggests that even if nature starts with a built in power law, we human beings can change those relationships over time if we have a clear motivation and common purpose with a systematic program of implementation. Just because the rich do get richer in a completely unregulated system, doesn't mean we cannot modify this natural, yet unhealthy tendency, through regulation and policies which even the playing field out a bit. In fact, this is exactly how Europe has managed to establish stable, just social democracies over the past sixty years, after the final breakdown of the combination of monarchies and dictatorships by the rich for the rich which had characterized Europe for many centuries.

Beyond Exteriors: Exploring our Interiors

Having spent 500 years perfecting modernism since the first Renaissance, more people alive today are experiencing levels of comfort and security than ever before in history, despite the brutally tough conditions of a billion or so in the developing world who have not yet benefitted from the material comforts of modernism. That is a not insignificant achievement of science and technology, together with the political and intellectual breakthroughs that yielded democratic systems of government for several billion people. At the same time, this system is built on a clockwork model of the universe in which mankind is lord and master of all he surveys, and nature is simply there to provide him with whatever he needs, without limit.

Completely separated from this miracle of science, technology and democratic mixed economies, we find belief systems of all kinds neatly sealed in millions of water tight compartments: religious believers, agnostics, atheists, mystics, spiritualists, complexity theorists—any and all attempts to explain the much bigger picture of which a tidy clockwork universe forms a very small part are all in there, pretty well sealed from the basic assumptions and mechanisms of modern life.

Of course, we cannot but help notice that our modernist, ethnocentric system is fraying around the edges, and slowly morphing into the equivalent of the butterfly emerging from the caterpillar cells in the chrysalis. While the caterpillar's business model is eternal growth:

simply eat as much as possible, for as long as possible, the butterfly that emerges through the imaginal cells dissolving the caterpillar's cells for its nourishment has a different goal: to explore and pollinate as many flowers as possible, while searching for food, a mate and a new home.

Our current political and economic systems are entirely focused on the caterpillar business model: eternal growth. There is not a politician in a major country who would ever speak out against economic growth as measured by GDP: gross domestic product, although several have ordered investigations into alternative measures of wellbeing such as GDH (gross domestic happiness).

Brutal definitions of success are more likely to produce a more brutal world. And yet, the US Declaration of Independence lists the "pursuit of happiness" as an "inalienable right" of mankind—rather ironically when viewed from the perspective of those who today see the USA as one of the most ruthless economic systems in the world. (There is no mention of the pursuit of GDP in the Declaration of Independence).

The idea that it is the business of governments to cheer up their citizens has moved in recent years to center-stage. Academics interested in measures of gross domestic happiness were once forced to turn to the esoteric example of Bhutan. Britain's Conservative-led government is compiling a national happiness index, and Nicolas Sarkozy, France's president from 2007-2012, considered replacing the traditional GDP count with a measure that takes into account subjective happiness levels and environmental sustainability.

The committee of economists, led by Joseph Stieglitz from the United States, reported back in 2009. President Sarkozy chose to receive—and endorse—the report one year to the day after the collapse of Lehman Brothers. The contradictory nature of his policies and personality, however, meant he blew his political capital long before he had a chance to take this promising idea any further.

At the Rio+20 Earth Summit, however, some of the world's leading economists presented the World Happiness Report, a 170 page tour de force which makes it clear that "Happiness" and "Wellbeing" will one day become priorities for governments, once their obsession with growth finally wears off. Here's to Gross Domestic Happiness—though perhaps we could lose the "Gross" part for once and for all, before economics students are forced to learn about GDH along with GDP.

While economists and politicians continue to debate GDP, GDH, MDG's and SDG's (Millennium and Sustainable Development goals), a shift is occurring in our social fabric worldwide as a result of the accelerated development the human race kick started with the first Renaissance.

The result of our steadily increasing ability to meet our basic needs as a species, together with a global revolution in education and training, means that every year more and more people are arriving at a stage in their lives where they can explore the interiors inside our modern exteriors. They do this both on their own and together in groups, in the "I" and "WE" spaces we described earlier. As this shift unfolds, a deepening and broadening of what it means to be human is occurring.

The simplest analogy to describe this deepening and broadening of human existence is the difference between two-dimensional, three dimensional and n-dimensional objects and experiences. As you read this book, you are interacting with a two-dimensional piece of paper with black ink printed on it, and your mind is supplying the third dimension by creating meaning from the flow of words you are reading so that what I am saying either makes sense to you, or it does not make any sense at all.

Yet this third dimension, which is supplied by your mind and your imagination, is in turn comprised of many more dimensions. One way of talking about dimensions in your mind is to describe them as contexts, processes and perspectives. For example, you might be reading these words at your desk in a noisy office, or in a comfortable armchair at home, or sitting at the top of a mountain in total solitude. The context you are reading this in includes not only the place you are in, but also the state you are in. You could be stressed, relaxed or blissed out, for example.

In turn, your state will determine what kinds of mental and emotional processes are available to you right now. If you are highly stressed, the presence of cortisol and adrenaline in your blood stream will ensure that your core flight, freeze or flight brain functions will be activated, and you will be highly alert to external stimuli, while your heartbeat will be raised and your blood will be flowing to you extremities to prepare you for action in response to a threat, demand or opportunity.

If you are relaxed in your armchair or blissed out on a mountain top, then your frontal cortex, the part of your brain that synthesizes your experience into meaningful wholes, will be activated. Your breath will likely be deeper, your heart rate slower and your ability to think conceptually and abstractly will be enhanced. In short, you are likely to be smarter and wiser in this state than if you were stressed out.

During the last century we have fleshed out the details of how we can change our state at will—something the sages have practiced down the ages. We learn to pay attention to our breath, center ourselves by taking a deep breath, and then empty our mind of distractions so that we can pay attention to what is going on in our interior. This process of centering is becoming increasingly wide spread, and leads to better decisions and outcomes for those who practice it, along with those who are affected by them.

Finally, depending upon the processes made available to you by your context, state and personal practices, you will become aware that you are operating in the moment according to one or more perspectives, which frame the situation you are in and what you believe your goals and options are right now. Even greater power and discernment now becomes available to those who can become aware of and master the art of perspective taking.

Earlier in this chapter we explored the eight primordial perspectives described by philosopher Ken Wilber in his integral approach. He calls these perspectives "primordial", because they are the most basic kinds of perspectives that can exist in the human mind, like quarks are the smallest particles physicists normally deal with outside of the particle accelerator.

So, all of our thoughts and feelings, no matter how complex or sophisticated, can be broken down into these eight basic perspectives. Even more usefully, we can use these perspectives in everyday life to examine situations from several angles to better understand our choices. For example, most of us will tend to have a favourite perspective at home, work, on holiday or at play. Many professions require their practitioners to operate out of exterior perspectives. For example, a medical doctor examines his patients from the perspective of the *individual exterior*, based on his knowledge of anatomy, physiology, biochemistry, and neurology and so on—all of which treat the human body and illness as purely physical phenomena.

An architect or town planner will also take an exterior perspective in designing a building or a new town—but this will be a *collective exterior* perspective because they must examine the interaction of the different parts of the systems they are designing to ensure that they meet the requirements of the design they are developing.

Other professions might require us to take more of an interior perspective on things. For example, both psychologists and psychiatrists explore the interiors of their clients, operating at the level of the *individual interior*. Artists, sociologists and literature professors and movie critics, however are required to understand, examine and work with the *collective interior* of our culture.

What is becoming increasingly clear in the 21st century, is that interior states and stages of development have powerful consequences personally and socially. Let us take the tragic example of the recent massacre of dozens of people at a midnight showing of the film "The Dark Knight Rises" in Denver, USA, by a schizophrenic dressed up as the villain "Bane", acting out a scene similar to that shown in the movie. There has been a media frenzy of analysis of the causes of this incident, as there have been in almost every one of the hundreds of such massacres perpetrated in the USA over the past few years.

Is it the presence of 300 million guns and poor gun control, is it the Wild West culture, is it violent movies, a failing mental healthcare system, poor security—the list goes on and on. The obvious answer is that it is all of these and more, but each "expert" has their favourite perspective from which to analyze the causes of the massacre, and "what should be done". And, it seems, symptomatic of their highly fragmented, materialistic culture, there is sufficient disagreement to ensure that more massacres of this kind will continue to be a sad feature of American life, as long as an integrated solution is not proposed and implemented.

We can scale up this problem to a global level. Whether it is Iraq, Afghanistan, Syria, Israel/Palestine, Chechnya, North Korea, Zimbabwe or Iran, the "problem" is viewed from different perspectives by the USA, Russia, China, and Europe and so on. Not only that, almost all of these "players" unaware that they are in the grip of their own perspectives, as they are submerged in their histories and cultures, usually completely hidden from view.

So, if a crazed PhD dropout can calmly shoot 12 people in cold blood because he believes he will be famous forever as the villain Bane, or Kim Il Jong, the "Great Leader" of North Korea thinks he needs to assert his power by launching a nuclear warhead at South Korea, then they are both right. And, given the current blindness we have built into our cultures and minds about our perspectives, whatever well-intentioned peace movements think or wish to believe, this will continue to be the case.

Our intentions drive our behaviors in many powerful ways, so that our minds create our individual and collective worlds. For human beings, consciousness involves experiencing the world as an extension of ourselves. This extension occurs through our expectations that we project onto the world, and our intentions through which we act in and on the world.

Every intention, and every expectation we have, all are constructed out of perspectives. For most people in the developed world, much of the time, the individual exterior perspective dominates. Westerners are largely focused on their material existence, as well as those emulating western lifestyles around the world. In many societies in the East, however, there are also a considerable number of people for whom the individual interior perspective is important, which changes the way priorities are set and lives are lived.

In both the East and the West, a much smaller number of people are engaged in working with collective perspectives, though this smaller number are generally more influential in their societies. Today in particular, those who have mastered the collective exterior perspective are essentially controlling the key systems on our planet that will determine whether we have a future as a species, and what kind of future that might be.

The Four Core Intelligences that Enable our Perspectives

Perspectives do not operate alone, however. Our minds are powerful, but they form part of a much deeper process that originates in our subconscious mind and emotions, which in turn are generated by our bodily processes and nature itself.

Many psychologists[5] use what is known as the "dual-process" model of learning, based on both our implicit associative processes or intuition

---

[5] For example, Daniel Kahneman Thinking, Fast and Slow, Farrar, Straus and Giroux, New York, 2011.

The Trouble with Paradise

(called "System 1"), and our more recent rational thinking processes or "reasoning" (called "System 2"). System 1 is responsible for our quick, intuitive thought processes, while System 2 is the source of slow, deliberative thought.

My own experience of fifty years of studying and practicing what I call "pragmatic psychology", has led me to add two even more crucial intelligences (also called "lines of development") to this list: System 0, our bodily or "kinaesthetic" intelligence and System 3, our transcendent or "spiritual" intelligence. There are other definitions of intelligence that break down the concept of intelligence into many more categories, such as Howard Gardner's theory of multiple intelligences[6], which is also very useful. For our purposes, however, the four main categories of intelligence we will examine are a "good-enough" starting point.

Moreover, our current understanding of ourselves implies the need for a more integrated approach to intelligence in general, and with the

---

[6] Psychologist Howard Gardner's theory of multiple intelligences breaks intelligence down further into several specific abilities in his 1983 book *Frames of Mind: The Theory of Multiple Intelligences*. He sets seven criteria for a behavior to be considered an intelligence: potential for brain isolation by brain damage, place in evolutionary history, presence of core operations, susceptibility to encoding (symbolic expression), a distinct developmental progression, the existence of savants, prodigies and other exceptional people, and support from experimental psychology and psychometric findings. Gardner chose eight abilities that he held to meet these criteria: musical-rhythmic, visual-spatial, verbal-linguistic, logical-mathematical, bodily-kinesthetic, interpersonal, intrapersonal, and naturalistic. He later suggested that existential and moral intelligence may also be worthy of inclusion. Although the distinction between intelligences has been set out in great detail, Gardner opposes the idea of labeling learners to a specific intelligence. Each individual possesses a unique blend of all the intelligences. Gardner firmly maintains that his theory of multiple intelligences should "empower learners", not restrict them to one modality of learning. He also argues that intelligence is categorized into three primary or overarching categories, those of which are formulated by the abilities. According to Gardner, intelligence is: 1) The ability to create an effective product or offer a service that is valued in a culture, 2) a set of skills that make it possible for a person to solve problems in life, and 3) the potential for finding or creating solutions for problems, which involves gathering new knowledge.

body being the source of all the rest of the "action" that comprises us human beings, our bodies prove to be a good starting point from which we can observe the emergence and inter-play of the other major systems of intelligence. We can regard Systems 0, 1, 2 and 3 as our "core operating system", with System 0 plugging into the "hardware" and System 3 both integrating all our intelligences while enabling us to plug into dimensions beyond our five senses.

System 0, our kinaesthetic intelligence, includes the ability to control our bodily motions and the capacity to handle objects skilfully. This also implies a sense of timing, a clear sense of the goal of a physical action, along with the ability to train responses. People who have bodily-kinaesthetic intelligence should learn better by involving muscular movement (e.g. getting up and moving around into the learning experience), and be generally good at physical activities such as sports, dance, acting, and making things. Our bodies are also the foundation for all our other intelligences; so developing our System 0 intelligence should also enhance all our other intelligences.

System 3 provides us with the ability to transcend and include Systems 0, 1 and 2 into a state of hyper aware mindfulness that gives us a choice between how we use our other three intelligences. System 3 is also our gateway to peak spiritual experiences, and enables us to view the world from a perspective of both detachment and compassion in the same moment. We are born with System 1 fully operational, and it takes another 12 to 15 years of formal education to perfect a reasonably effective System 2 intelligence.

System 3 may develop alongside System 2, but in some people and cultures it may be regarded as slightly alien, particularly in the most hyper-rational, materialist cultures, where it is usually separated off in an unexamined watertight compartment called "religion" or even "atheism". It is interesting that even in officially atheist communist systems such as Soviet Russia and Maoist China, this sense of "a bigger picture", overview perspective, remained alive in those cultures despite several decades of official suppression. Immediately such official suppression was lifted, both religious and atheistic forms of spirituality returned in full force in large proportions of the population.

System 1 is automatic and unconscious, based upon our experience, and exhibits the universal cognition that is shared between humans and animals. System 1 controls instinctive behaviours and tends to solve problems by relying on prior knowledge and belief. Rapid, parallel and automatic processes characterize System 1 thinking where only the final product is conscious.

System 2, our general purpose reasoning system, evolved late in comparison with our intuition. System 2 works alongside the older autonomous sub-systems of System 1, and our success as a species appears to be based largely on such higher cognitive abilities. Evolutionary theorists believe that System 2 emerged around 50,000 years ago when representational art, imagery, and the design of tools and artefacts are first documented, giving rise to our logical, reasoning mind.

System 2 is also known as the explicit system, the rule-based system, the rational system, or the analytic system. It performs slower sequential thinking. It is domain-general, performed in the central working memory system. Because of this, it has a limited capacity and is slower than System 1, but enables us to think abstractly and hypothetically.

To survive physically or psychologically, we sometimes need to react automatically to a speeding taxi as we step off the curb or to the subtle facial cues of an angry boss. That automatic mode of thinking, not under voluntary control, contrasts with the need to slow down and deliberately fiddle with pencil and paper when working through an algebra problem. System 1 loves to make stories and jump the gun, while System 2 is cool-headed reason. It is System 2's task to put the reins on the mind's Don Quixote when he is about to charge at a non-existent windmill.

When we think of ourselves, we identify with System 2, the conscious, reasoning self that has beliefs, makes choices, and decides what to think about and what to do. Although System 2 believes itself to be where the action is, automatic System 1 effortlessly generates impressions and feelings that are the main sources of the explicit beliefs and deliberate choices of System 2. The automatic operations of System 1 generate surprisingly complex patterns of ideas, but only our slower, more deliberate System 2 can construct thoughts in a logical series of steps. In certain situations, it is also vital that System 2 takes over and overrules the freewheeling impulses and associations of System 1.

In his most recent book, "Thinking, Fast and Slow", psychologist Daniel Kahneman shares with us the drama of what can go wrong when System 2 is too slow to keep System 1's impetuosity in check. Our intuitive mind (System 1) is capable of amazingly fast, and often correct, hunches, as Malcolm Gladwell spends most of the pages of his very popular book "Blink" telling us. Yet, there are serious limitations on the kinds of problems System 1 is good at solving, which lead us to make some very bad decisions and reach some very wrong conclusions. In fact, the more complex the world becomes, the more likely System 1 is to screw things up.

We are therefore guilty of unintentionally creating many little Hells on Earth through our natural tendency to give System 1 free rein—from stock market and economic crashes to marriage breakups to actions that land us in the criminal justice system, we find that our instincts and our hunches are often badly wrong with hindsight. System 1 suffers from being over confident, often rushes in where angels fear to tread, and tends to see what it wants to see in a situation. Yet it also gives us a huge amount of pleasure, and a heightened sense of being alive, so those who are addicted to System 1 thinking are often the last to find out that they have become a walking disaster zone when they've made one too many seriously bad decisions.

Au contraire, hyper rational Dr Spock characters may be frequently right using their System 2 reasoning capabilities, but they can also turn out to be a real yawn at dinner parties. Accountants, statisticians, physicists, actuaries, mathematicians, computer programmers and other ultra—logical people can often live in a world of their own. Clearly some kind of balance between our System 1 and System 2 modes is required for a satisfying existence, and this is precisely the function of System 3.

Like meditation, mindfulness has become very popular in the past few decades amongst the business and ruling classes, while also being essential for new age and postmodern seekers. The benefits of mindfulness appear to be that one can become mindful of both System 1 and 2 at the same time, and consciously use both modes as tools of thought rather than see them as something fixed and immutable. In fact, simply being aware of your own System 1 and 2 in operation, would give you more of an opportunity to think through and reflect on your own being and doing.

This also implies a full awareness of our bodies, and all the significant neural processing power that drives our hearts, guts and other vital organs. For many decades scientists have also been able to measure the human energy field that surrounds our bodies, and the interaction between our body minds and our energy fields is now well documented. Each of us is ultimately capable of being aware of and to some extent influencing our entire presence from our individual organ systems to our energy fields and beyond, though this does represent a significant departure from the traditional medical paradigm which prefers to break our bodies down into parts and work individually with each system, shunting us from one specialist to another until they can diagnose "the problem", as if we were a motor car.

Taken at its fullest expression, mindfulness then becomes a way of being aware in every moment of each and every nuance of yourself, as sensed through every facet of your being. You are totally in the "Now", in a flow state where your attention is able to permeate your own interior, through your awareness of your exterior, while also being fully cognizant of the people and world around you, and their interiors. Such genius or saint-like qualities are indeed still rather rare amongst our species, but we find them more often around the world as our evolution accelerates.

From such awareness we sense what the Anglo-Indian sage Krishnamurti pointed out, which is the place of the ever present moment between thought and action, from where we become more powerful in making better and wiser choices for ourselves and others. We might say that System 0 (our physical, energetic and cultural context) is "live streaming" through our intuitive System 1 and rational System 2, and that we are observing all three systems from our mindful System 3, which is what the Buddhists have referred to for three millennia as non-dual consciousness—the Witness which sees everything else, but itself cannot be seen or described except as "that".

It would be from this place of pure unfiltered awareness that we could see the truth that what we can imagine we can will, and what we will can eventually become our reality. Becoming aware of our true desires, we are then able to start creating the life of our dreams, even if the world insists upon intervening rudely from time to time with setbacks to remind us that we are still in need of new learning and transformation ourselves in that very process.

If we become totally free of all perspectives and thoughts, would we not still be in the grip of our unconscious values and other hidden assumptions about our world that we do not even know we have, and that effectively "have us"? Every "I" is part of a collective "WE" and

"ITS", so that what we are seeing in our witnessing state of mindfulness will still be interpreted by us from our existing world view, based on the stage of development we are at, and the values implicit therein. Thus one man's heaven may be another man's hell.

The Paradise Creation Business is a Multi-Level Enterprise

Humankind is in the paradise creation business. Just as naturally as a cat always finds the perfect spot on the rug in front of the warm fire to curl up on of a cold winter's night, so too does every human being seek their perfect spot. Until 500 years ago, however, occupying the human niche was characterized more by demanding life conditions than it was by pleasure seeking and personal growth, combining a great deal of suffering with some small measure of enjoyment from time to time.

It is understandable that given the difficulty of creating heaven on earth, we created a heaven in the sky with the gods of our choice, to comfort and console us in our times of need—a place completely untouched and untouchable by the hardships we experienced on a daily basis here on earth. The Hindu concept of Moksha and the Buddhist concept of Nirvana may be thought of as a kind of utopia. In Hinduism or Buddhism, however, Utopia is not a place but a state of mind—a belief that if we are able to practice meditation without a continuous stream of thoughts, we are able to reach enlightenment. This enlightenment promises exit from the cycle of life and death, relating back to the concept of utopia.

During the past five centuries, it has become possible to dream of, and even attempt to build, Heaven on Earth. In 1516 Thomas More wrote his now famous book "Utopia", which proposed an ideal society. No doubt he had read some of Plato's ideas on the subject, though Plato's "Republic" categorized citizens into gold, silver, bronze and iron socioeconomic classes, with the gold class acting as the rulers and philosopher kings. In Plato's <u>Utopia</u> there are no laws or lawyers, and the Republic defends itself with mercenary armies who can help subdue warlike tribes into peace.

The Trouble with Paradise

Since More's Utopia, philosophers and activists have designed and built many different versions of Utopia. Early America was built by such idealists from the Quaker, Shaker and other persuasions, and the American constitution was drafted by these idealists, combining the best of the British, French and other systems of government. There are ecological utopias, feminist utopias, economic utopias, religious utopias, science fiction utopias and more. Buckminster Fuller presented a theoretical basis for technological utopianism and set out to develop a variety of technologies ranging from maps to designs for cars and houses which might lead to the development of such a utopia.

What all of these attempts at the creation of a Utopia have in common, is that with a few, small rare exceptions, none of them have remained faithful to their original design, or survived the ravages of time.

What is also certain is that one man (or woman's) Utopia, is another's Dystopia. Not surprisingly, our conceptions of our own personal paradise arise from the stage of development we are at, and we are all at different stages of development, from our infancy to our old age. And each society also evidences its own unique series of stages of development, though there are similarities between many of them.

What has emerged globally over the past five centuries is a system of world trade between nation states dominated by markets, businesses and governments, with technological progress at the center of the whole system. Ideally, this system lets the individual decide how they want to live, what they want to do to make a living, and provides them with the resources to get on and do just that. In such a system we make our progress in the world on the basis of our merit, and our ability to apply technology to add value or be of service to others. In other words, a technocratic meritocracy, on a good day, where each person has the freedom of choice to create their own life in their own way, believing whatever it is they wish to believe.

In the 21st century we are living in a time where most of the developed world has embraced a form of capitalism, ranging from the egalitarian social democracies of northern Europe to the semi-feudal dictatorships found in many developing economies. Capitalism is a decidedly non-Utopian system, which at its roots focuses primarily on economic growth and material wellbeing. Markets have no ethics or morals—only people do, and they differ widely in what they believe to be right and wrong.

Until the 20th century and the rise of psychology in the popular consciousness, no overarching model existed that could explain how it was that so many people could believe they alone were right, and that everyone else was wrong. The 20th century itself was characterized by clashing ideologies from capitalism to communism, from militaristic fascists to peace-loving hippies, and hundreds of millions of people died untimely deaths in the name of those ideologies. During the 21st century we have an opportunity to transcend these polar opposites, to move toward the syntheses of the clashing theses and antitheses so beloved of mankind. But how so?

There are two kinds of understanding which are becoming increasingly available to those who care about creating a flourishing world in the 21st century, rather than heading down into an apocalypse. The first is a form of intelligent empathy[7] that enables us to connect with others no matter whom they are or where they come from. This empathic intelligence is now being cultivated from a very young age in major cities and places where the full diversity of mankind is coming together from nursery school through university and beyond into the workplace.

Such intelligent empathy is based both on a much more comprehensive awareness of the breadth and depth of the human experience around the planet, as well as direct experience of people from other cultures and races. At its best, this empathic intelligence also extends to nature and other species, though that is sometimes a real challenge for the increasingly urbanized populations of the world, whose experience of nature in the wild is increasingly limited.

---

[7] For an overview of the literature on empathy see: Simon Baron-Cohen (2003): Empathy is about spontaneously and naturally tuning into the other person's thoughts and feelings, whatever these might be. There are two major elements to empathy. The first is the cognitive component: Understanding the others feelings and the ability to take their perspective . . . the second element to empathy is the affective component. This is an observer's appropriate emotional response to another person's emotional state. The Essential Difference: The Truth about the Male and Female Brain, Basic Books (July 1, 2003)

The second kind of understanding that is facilitating our flourishing is a version of Big History[8], which is a field of study that examines history from the beginning of time to the present day, focusing on both the history of the non-human world and major adaptations and alterations in human experience. While Big History as a field focuses mainly on our collective exterior, Big Psychology focuses on both our individual and collective interiors as well. Such an understanding of ourselves provides us with a better understanding and appreciation of both the nature and motivations of others, as well as insights into our own nature and development.

There are a few hundred different versions of Big Psychology, each with its own particular vocabulary and biases. What is remarkable, however, is the extent to which all of these developmental models settle on eight or so levels of human development, irrespective of their origins. Our ability to see these different levels of development in our own personal history, and ourselves, provides us with a greater capacity to relate to and empathize with the place others are at, and where they are coming from.

Big Psychology also enables us to not only appreciate where people have come from and where they are today, but also what might be next for them in their own development and life journey. Of course, as with any science, we have to be careful that we do not use our knowledge to manipulate people or situations for our own ends in unethical ways, although this is probably inevitable with most human beings who would like to see one set of outcomes preferred over another. Even if you cannot use this knowledge completely objectively, you are honour bound to at least use it for the highest possible good for as many people as possible. Do I have your word on this?

---

[8] Take a look when you can at a fabulous online map of Big History called ChronoZoom, put together by Bill Gates and a variety of educationalists. You can explore Cosmos, Earth, Life, Human Prehistory and Humanity interactively through the internet: http://www.chronozoomproject.org

Chapter Seven

The Arc of Human Development in the 21st Century

The present moment finds our society attempting to negotiate the most difficult, but at the same time the most exciting transition the human race has faced to date. It is not merely a transition to a new level of existence but the start of a new 'movement' in the symphony of human history".

Psychologist Professor Clare Graves: Speech to the American Psychological Association 1970

In previous chapters we peeked inside Big History to illuminate the trajectory of our species including our 15 million year journey from large apes to homo sapiens sapiens, the wise naked ape, and learned that our evolution, just like that of the natural world, proceeds in the direction of more complex, collaborative and open systems. The same principle applies to the development of our consciousness and culture, where our individual and collective interiors evolve. Big Psychology makes it clear that psychological and cultural development follows a pattern, and that pattern is always from more partial to more whole[9].

In the 21st century not only can we expect a continuation of this arc of development: we can also expect a great leap forward in human civilization, as we move through the conventional stages of development to a new phase in human history characterized by empathic, integrated ways of being and doing that are global in perspective. Yet we are also

---

[9] Since research began in the mid-20th century eminent scholars and psychologists have independently discovered similar stages of human psychological development. Building on the shoulders of giants such as Freud, Piaget, Kohlberg, Graves and Maslow, the latest research by Kegan, Loevinger, Wade, Beck, Cowan and Wilber confirms the five to seven developmental levels in the original models while extending these models with a further two to three levels.

in a life and death struggle between the new consciousness, culture and systems that are emerging and the old ways of "business as usual" which is fighting what is emerging with great cunning, deception and massive amounts of money.

All around us we can see that this great shift is a monumental struggle, between old and new ways of thinking and doing, between the "haves" and "have not's" in the old system and the new forms of wealth being created, and between those who have power in the old systems and those to whom power is shifting as the new systems gradually take hold. Yet evolution is an unstoppable force, and it has a clear direction, which in the case of human development, is "UP", even while the reactionaries are furiously stabbing at the down button in the elevator.

So what does "UP" look like, and how might we get there from here? What obstacles might lie in our path? And what would this mean for each of us personally; for our community; and for our world?

Evolution in the Direction of "UP"

Both evolutionary and revolutionary processes drive the trajectory of human development. Evolution is slowly becoming a conscious collaborative process, driven by our hearts and minds, rather than a blind, unconscious and competitive struggle driven by unseen forces out of our control. The human species is maturing from a troublesome teenager to a more awake and progressive twenty-something, though our progress is measured in centuries rather than years.

The revolutionary aspect of our development lies in our ability to transform ourselves and our systems and technologies, making predictions into the future difficult if not impossible in most areas of life. Whether these transformations are political (from the fall of the Berlin Wall to the Arab Spring), technical (from a million songs on your iPod to cures for most major diseases), or social (think of the rise of social media and virtual communities in less than a decade), various aspects of our lives are undergoing radical transformations every decade. Imagine we find solutions to overpopulation (the global birth rate is already falling below UN estimates), and global warming/

climate change in the next decade? What is hard to believe right now can become commonplace in a decade—that is the lesson of the last century.

So it is that we find the messy process of evolution delivering greater complexity and consciousness over the longer term, while being punctuated by radical breakthroughs or breakdowns from time to time. Within the general trend of "UP", we find both major progressions and major regressions, with the latter often acting as fuel for the former, embodied in the formula: "What does not grow, decays". In the human body we find the rate of cell growth and cell death sufficiently evenly balanced so that we do not notice that our entire body is renewed on average every seven years. In stable societies and civilizations, we also find a similar ratio of growth and decay in the population and the artifacts and infrastructure they live in.

What drives the evolution of human beings is the interplay between the growth and development of individuals, and their environment. Historically our environment dictated our life conditions, and it is only relatively recently (in the past few millennia), that we have been able to modify our environment to any significant extent. Scientists and historians are sufficiently impressed by our 21st century environment modifying abilities to name the geological age we are in the "Anthropocene"[10].

The interplay between our environment and ourselves is mirrored in us by the interplay between our exteriors and interiors—our bodies and our minds. From an individual perspective we are able to track three key developmental stages. In the first two stages, pre-personal and personal, we are operating out of the reptilian brain and the limbic system respectively.[xlix] The R-complex and limbic systems effectively correspond to System 0 (the core bodily functions), and System 1

---

[10] Many scientists are now using the term and the Geological Society of America titled its 2011 annual meeting: Archean to Anthropocene: The past is the key to the future. The Anthropocene has no precise start date, but based on atmospheric evidence may be considered to start with the Industrial Revolution (late 18th century). Other scientists link it to earlier events, such as the rise of agriculture. Evidence of relative human impact such as the growing human influence on land use, ecosystems, biodiversity and species extinction is controversial, some scientists believe the human impact has significantly changed (or halted) the growth of biodiversity

(the intuitive, analog intelligence which has often been called "right-brained"), as we saw earlier in chapter 6 above.

The bridge from the personal to the transpersonal phases of development is the neocortex, which in man constitutes two-thirds of our brain. A person without a neocortex is essentially a "vegetable" as they kindly put it in the medical business, while a mouse without a neocortex can function normally for all intensive purposes, so the development of the functions of the neocortex is crucial for humans to evolve into responsible, conscious, choice-making citizens in the highly complex world we have evolved over the past few thousand years.

We find System 2, the digital, procedural intelligence, emerging in this collaboration between System 0 and System 1, enabling us to program our actions and behaviours much more effectively, in the pursuit of specific goals, and in collaboration with others.

In the transpersonal phase of development the pre-frontal and frontal cortices become new control centres which enable individuals to activate System 3, the "strategic psychological helicopter", from which they are capable of getting an overview of the situation they are in and in which they are capable of being mindful and taking better decisions that benefit the greatest span of human beings while honouring the greatest depth in those individuals.

Below is a diagram of the "triune brain" proposed by neurologist Paul MacLean in 1952.

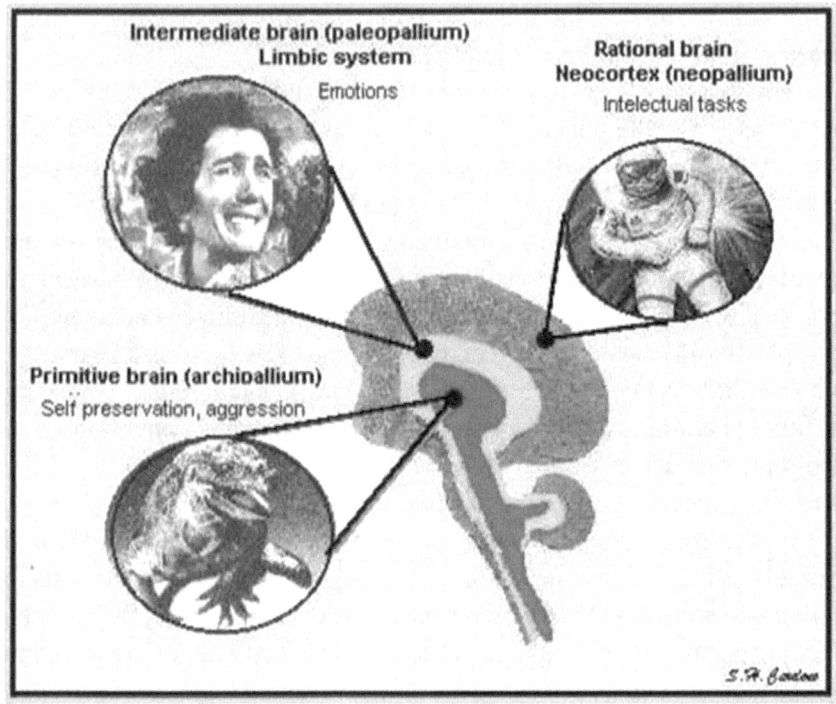

**Diagram 1—The Triune Brain Model
of Neurologist Paul MacLean**

1. **Pre-personal Phase**—in our first decade or so of life we develop through the pre-personal phase. In the developed world this stage results in the emergence of a self that is distinct from our family, and in the developing world a self that is distinct from our tribe. Our heroes during this stage of development are powerful, often impulsive, egocentric and, of course, heroic. During the pre-personal phase we believe in Santa Claus, magical-mythic spirits, dragons, beasts, and powerful people. In ancient times these were archetypal gods and goddesses, while in the 21st century we find super-heroes and their archenemies embodied in books and film, from Harry Potter to Lord of the Rings to Batman. While we are developing our rational faculties during this phase, unconscious motivations and magical thinking often characterize our behavior.

Dysfunction during these years can lead to issues arising throughout a lifetime. Challenging or primitive life conditions can also make it difficult for people to move beyond these developmental stages, and during times of war, famine, great hardship or personal difficulty, people often regress to these stages as mature adults. In such cases the family or the tribe is the ultimate shock absorber.

2. **Personal Phase**—At some point between being a child and becoming a teenager, we learn to be conscientious, responsible people, if all goes well, and experience a desire to conform to conventional norms and behaviours. With the emergence of our own conscious identity and rational mental faculties, we are now able to strike a balance between our intuitive (System 1) and reasoning thought (System 2) processes at a conscious level. We may even be experiencing some System 3 moments where we transcend ourselves and the situations we are in, gaining a helicopter perspective and perhaps even a "peak experience" with spiritual Aha! Moments.

    Modern educational systems, modern organizations and institutions and some religions are powerful forces in helping to shape such conventional and rational mental processes. Those who have been shaped by such systems often reach the achievement and Affiliative levels by their mid-twenties and form the majority of the population in most developed societies. Modern "civil" society is based upon conscious mental processes being activated and used on a daily basis by most people.

3. **Transpersonal Phase**—In the transpersonal realm one transcends and integrates all other developmental levels, moving through the authentic integral level to the transcendent and unity levels of higher consciousness. According to recent research, hundreds of millions of people are now actively exploring these transpersonal levels worldwide, while at least 1% of the world's population is anchored at the transpersonal in their daily life.

    This is the great leap psychologist Clare Graves was talking about in 1970 as he reviewed his latest research results with Abraham Maslow and the rest of the American Psychological Association members. Such post-modern stages of development are the basis upon which an integral, global civilization could be built in the 21st century. I say could because:

- the billions of conformist and achievement oriented power holders would have to be sufficiently attracted to the possibilities that they are able to let go of some of their narrow belief systems and vested interests to give the newer systems and structures room to grow;
- the hundreds of millions of affiliative, cultural creatives would need to become much more grounded and practical in their desire for transformation and demands for change, while also shedding the last remnants of their often narcissistic tendencies.

Diagram 2 below offers a simplified representation of the different pathways the evolution of human consciousness can take. This model simplifies the three main, overlapping models of human development currently used in a wide variety of organizations, psychology and philosophy textbooks. The names of the successive stages follow the classification system used by psychologist Jenny Wade and the colors used by Professor Don Beck, the co-creator of Spiral Dynamics.

The "Reactive" (or "Beige") stage evolved about 100 000 years ago when our distinct sense of self began to emerge, and food, water, warmth, sex and safety were our top priorities. We used our very wide range of senses (far wider than the five main senses used today), instincts and habits to survive in the wilderness, and formed into small survival bands to perpetuate life, living off the land close to nature along with the other animals. The original Bushmen of the Kalahari Desert, aboriginal tribes on various continents including Australia and the remaining tribes of the rainforests in the Amazon, Borneo and elsewhere living in small bands of up around 30 people are typical of this stage of development, where the most highly developed part of the brain is the R-complex.

Around 50 000 years ago, a new stage of development began to emerge when we started to invent art, music and religion—in short, the beginnings of culture and metaphorical thinking that transformed our ability to communicate and coordinate ourselves in larger groups ranging up to 300 people. This new tribal form of existence placed a priority on the individual being a loyal group member, showing allegiance to the tribal chiefs, elders, ancestors and the clan.

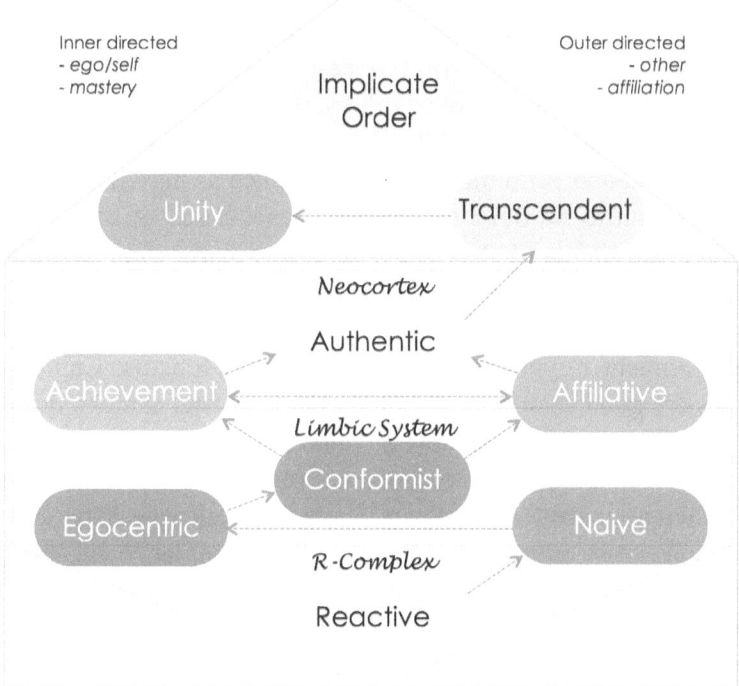

Diagram 2—9 Stages in the Evolution of Human Consciousness

At this Naïve (or "Purple") stage of development, sacred objects, places, events and memories are paramount to making sense of the world. Observing the rites of passage, seasonal cycles and tribal customs are essential for tribal cohesion and bonding between its members. Spirit beings and mystical signs are paid close attention to, ensuring that individuals have a sense of belonging and safety in the face of an unpredictable and often cruel world. People at this stage of development are outer-directed towards others, and value their affiliation with the group above all else. The 400 or so original tribes of Europe and the 400 or so original tribes of Africa were all interconnected through a variety of familial and trading alliances, yet also engaged in warlike behaviour frequently when life conditions got tough or over-crowded.

About 10 000 years ago, the egocentric (or "red") stage of development began to emerge. Like the magical/naïve purple stage of development, this stage is a halfway house between the R-complex part of our brains and the limbic system, which processes emotions and stores a treasure house of memories. Unlike the magical/naïve purple stage, however, this egocentric red stage is driven by a focus on self and the mastery of power over others. Through our development of a variety of technologies 10 000 years ago, the world became at once a more settled and yet a more dangerous place as massive armies and empires began to be built for the first time, based on the surpluses generated through settled urban populations harnessing a variety of agricultural, building and military technologies, as well as elaborate social hierarchies.

The ancient Egyptians, Greeks, Persians, Mongols, Chinese and Romans are good examples of Empire builders who came to dominate parts of Europe and Asia for several hundred years at a time, only to be pushed back to their native territories as other more powerful Empires rose in their ascendant. In South and Central America the Aztec and Inca Empires also rose more recently and exhibited similar social structures and power dynamics, though with very different life conditions and cultural content.

Egocentric, "Red" stands tall, expects attention, demands respect and calls the shots, and attempts to dominate competitors by conquering or out-foxing them. Being what you are having what you want, regardless of the consequences for others, is the primary goal of such individuals and cultures. The emphasis of life is to enjoy oneself to the fullest right now without guilt or remorse, and breaking free from any domination or constraints as required to fulfil immediate desires. Egocentric red is thus highly impulsive, generally opportunistic, and often unpredictable.

About 5 000 years ago, major religions began to arise and spread around the planet, beginning with Egyptian and Chinese religions. During the Axial Age which began around 3 000 BCE and ended around 700 CE, an intense burst of new religion creation began starting with Buddhist and Hindu traditions, followed by Judaism, Christianity and Islam. Over the course of these 6 000 years, a new way of being emerged centered around a variety of belief systems encapsulated in "new stories", that gave life a meaning, direction and purpose with pre-determined outcomes. It must be stressed that not only are these

The Trouble with Paradise

religious belief systems—they also represent the emergence of new capacities in the human brain and being to recognize and participate in new forms of order that also have consequences for governance systems and culture, along with technological and scientific advances appropriate to that level of development.

As we saw earlier in chapter 2, each religion had its founding leaders and/or gods, such as:

- Buddha and his three universal truths (Buddhist have no gods),
- The Greek, Persian and Roman gods gained popularity along with the social stability imposed by their Empires and relatively advanced administrative, political and legal systems
- the Egyptian gods Ra and Osiris, always linked into Egyptian royalty and priests
- the three principal gods of Hinduism—Shiva, Vishnu and Shakti,
- Moses and his God Yahweh, followed by the Hebraic Kings and Prophets and
- Jesus and Mohammed, both believed to be sons of God and god-like by Christians and Muslims respectively.

Such Conformist "True Blue" belief systems require a degree of sacrifice to their specific transcendent Cause, Truth or Righteous Pathway. There is a strong hierarchical order in which the powerful enforce a code of conduct based on eternal, absolute principles that, if followed and obeyed, produce stability now and guarantee future reward, whether in heaven or on earth. The impulsivity of the egocentric red way of being is now controlled through guilt and a desire to "do the right thing", where there is a place for everyone, and everyone is in their place (though not necessarily wildly happy about it). Laws, regulations and discipline build character and moral fiber, such that social and economic arrangements become more certain, predictable and enforceable.

The conformist stage of development is the last stage centered predominantly in the limbic system. Depending upon an individual's predisposition, the next stage of development can either be achievement or affiliation oriented. Some individuals might be balanced evenly between these developmental stages, though women tend to develop more strongly in the direction of affiliation, and men tend to develop more strongly in the direction of achievement.

The shift from the limbic system to the neocortex as the control center that determines our priorities and creates our options and actions is a major advancement in human civilization. This was precisely the shift that occurred during the first European Renaissance starting in about 1500 in northern Italy. It has taken fully 500 years for the "Achiever" and "Affiliator" modes of consciousness to mature into fully-fledged ways of operating that form the center of gravity of our modern and post-modern cultures in the developed world.

The Achievist ("Orange") level of development began to put down deep roots in European circles about 300 years ago. At its root this kind of consciousness is optimistic, risk-taking and stresses science, technology and competitiveness. It acknowledges that change and progress are built into the nature of things, and focuses on learning the secrets of nature and implementing the best solutions to hitherto difficult challenges. This modernist approach manipulates the earth's resources to create and spread abundance, without recognizing limits to growth. Achievist strategies are built around acting in one's own self-interest by playing the game to win—an approach Adam Smith would have applauded and which he set out clearly in his book "The Wealth of Nations" a few centuries ago.

It is no surprise, therefore, to learn that the Affiliative level of development exists in a strong tension with the Achievist level of development. This communitarian/egalitarian ("Green") mode of consciousness emerged slightly later than and often in reaction to aspects of the achievist "orange" system. It can be seen in the better parts of the French Revolution, the American constitution, the early Quakers, the movement for the abolition of slavery, and various European philosophers of the eighteenth and nineteenth centuries. Where the achievist level seeks to create wealth, the affiliative system is concerned with the equitable distribution of wealth, and is associated with Socialist and other liberal movements over the centuries.

The focus of green affiliators is on seeking peace within and without, co-creating caring in communities where feelings and sensitivity supersede cold rationality. One of the major contributions of "green" consciousness and movements it that it redresses the imbalances created by harsh blue TruthForce systems and often too greedy orange Achievist systems, helping heal old wounds and bringing reconciliation, truth and justice to the fore. The affiliative green system was at the forefront of the South African peace and reconciliation process after the harsh years of apartheid, enabling Nelson Mandela to be the "good guy" who took care of all South Africans, not just his own supporters.

The six levels of development from the Reactive Beige system through Magical Tribal Purple, Egocentric Red, Conformist Blue, Achievist Orange and Affiliative Green, are what is known as "First-Tier" systems. This is because they all share the common characteristic of believing that they are each "the only game in town". As one ascends from Beige through to Green, the higher levels systems take their own superiority for granted, as they are more complex and extend further in space and time than each of the previous levels. Blue TruthForce believes that if it could only evangelise everyone else effectively, then the "others" would sign up to their belief system or culture. Orange Achievists believe that everyone secretly wants progress above all else, and that people will do anything to get ahead. Green Affiliators believe that orange achievists are greedy and overbearing, and need to see the error of their ways, while those Blue Conformists are just so stuck in the mud and backward.

In reality, however, each value system is an adaptation to a specific niche, and in its own ways will be better or worse adapted depending upon specific life conditions prevailing in specific niches at certain points in time. The extent to which there is "dissonance" between people's values and their life conditions will determine to what extent their dissatisfaction with what no longer works for them will translate into action that leads them to evolve a more appropriate blend of values for their needs.

With the completion of the "Green Meme", human consciousness is poised for a quantum jump into "Second-Tier thinking." Clare Graves referred to this as a "momentous leap," where "a chasm of unbelievable depth of meaning is crossed." In essence, with second-tier consciousness, one can think both vertically and horizontally, using both hierarchies and heterarchies (both ranking and linking). One can therefore, for the first time, vividly grasp the entire spectrum of development, and thus see that each level is critical for the health of the overall Spiral. The Authentic, Integral, level of development is the starting point for this exciting journey, followed by the Transcendent and Unity stages of development. The latter two stages are not well mapped yet, as there are so few people in them at the moment.

Each wave of development transcends its predecessor, and yet it includes or embraces it in its own makeup. For example, a cell transcends but includes molecules, which transcend but include atoms. To say that a molecule goes beyond an atom is not to say that molecules hate atoms, but that they love them: they embrace them in their own makeup; they include them, they don't marginalize them. Each wave of existence is a fundamental ingredient of all subsequent waves, and thus each is to be cherished and embraced.

Moreover, each wave can itself be activated or reactivated as life circumstances warrant. In emergency situations, we can activate red power drives; in response to chaos, we might need to activate blue order; in looking for a new job, we might need orange achievement drives; in marriage and with friends, close green bonding. All of these value systems have something important to contribute.

But what none of the first-tier value systems can do, on their own, is fully appreciate the existence of the other value systems. Each of the first-tier value systems thinks that its worldview is the correct or best perspective. It reacts negatively if challenged. Blue order is very uncomfortable with both red impulsiveness and orange individualism. Orange individualism thinks blue order is for suckers and green egalitarianism is weak and woo-woo. Green egalitarianism cannot easily abide excellence and value rankings, big pictures, hierarchies, or anything that appears authoritarian and thus green reacts strongly to blue, orange, and anything post-green.

All of that begins to change with second-tier thinking. Because second-tier consciousness is fully aware of the interior stages of development—even if it cannot articulate them in a technical fashion—it steps back and grasps the big picture, and thus second-tier thinking appreciates the necessary role that all of the various value systems play. Second-tier awareness thinks in terms of the overall spiral of existence, and not merely in the terms of any one level.

Where the green value system begins to grasp the numerous different systems and pluralistic contexts that exist in different cultures (which is why it is indeed the sensitive self, i.e., sensitive to the marginalization of others), second-tier thinking goes one step further. It looks for the rich contexts that link and join these pluralistic systems, and thus it takes these separate systems and begins to embrace, include, and integrate them into holistic spirals and integral meshworks. Second-tier thinking, in other words, is instrumental in moving from relativism to holism, or from pluralism to integralism.

The Integral Stage of Development—MindShift and WorldShift

The European philosophers Jean Gebser and Ervin Laszlo, and American philosopher Ken Wilber have written a great deal about what they call the integral stage of development—Wilber to the extent that he calls his life's work "Integral Philosophy". If Wilber and others such as Paul Ray, researcher and author of the "Cultural Creatives" are right, then roughly one hundred million plus individuals are engaged in a transition from first-tier ways of being and thinking, into second tier or "Integral" ways of being and thinking. What would that mean for the evolutionary trajectory of our species Homo sapiens, and what would that mean for each of us who are aware of this approach?

Earlier in this book we briefly explored Wilber's Integral Theory which places perspectives at the centre of his way of describing our world and Kosmos. Psychologist Don Beck has also explored this developmental level using colourful terminology for the first stages of integral development (which he calls "Yellow", and then "Turquoise" for the more holistic, planetary perspective). Wilber makes some grand claims for his integral approach—here is the introductory page on Amazon for his book "Integral Vision":

"Suppose we took everything that all the various world cultures have to tell us about human potential—about psychological, spiritual, and social growth—and identified the basic patterns that connect these pieces of knowledge. What if we attempted to create an all-inclusive map that touches the most important factors from all of the world's great traditions?

Ken Wilber's Integral Vision provides such a map. Using all the known systems and models of human growth—from the ancient sages to the latest breakthroughs in cognitive science—it distills their major components into five simple elements, and, moreover, ones that readers can verify in their own experience right now.

In any field of interest, such as business, law, science, psychology, health, art, or everyday living and learning—the Integral Vision ensures that we are utilizing the full range of resources for the situation, leading to a greater likelihood of success and fulfillment. With easily understood explanations, exercises, and familiar examples, The Integral Vision shows how we can accelerate growth and development to higher, wider, deeper ways of being, embodied in self, shared in community, and connected to the planet, which can literally help with everything from spiritual enlightenment to business success to personal relationships."

We do indeed live in an extraordinary age, where all of the world's knowledge and cultures, are available to each of us—something unprecedented in the history of our species. For the past few hundred thousand years a person was born into a culture that knew only of its own existence. For example, someone born in Africa, was raised as an African, married an African, and followed an African religion—often living in the same hut for their entire life, on a spot of land that their ancestors settled millennia ago. We have evolved out of Africa from isolated hunter gatherer bands like the Bushmen, to the tribal settlements and farms of the Xhosa, to the ancient nations of Africa, to the conquering feudal empires of the Zulus and other warring tribes. In the past five hundred years we have also witnessed the emergence of international corporate states such as the Dutch East India Company to modern multinationals such as Unilever, and the other inhabitants of the global village including the United Nations and hundreds of thousands of NGO's and charities. We are now witnessing the next stage in our evolution toward an integral global village that seems to be humanity's destiny.

As Wilber puts it: "So it is that the leading edge of consciousness evolution stands today on the brink of an integral millennium—or at least the possibility of an integral millennium, where the sum total of extant human knowledge, wisdom, and technology is available to all. But there are several obstacles to that integral embrace, even in the most developed populations. Moreover, there is the more typical or average mode of consciousness, which is far from integral anything, and is in desperate need of its own tending. Both of those pressing issues (the integral vision as it relates to the most developed and the least developed populations) are related directly to the contents of this volume of the Collected Works."

Ervin László's Integral Theory of Everything [11] posits a field of information as the substance of the cosmos. Using the Sanskrit and Vedic term for "space", Akasha, he calls this information field the "Akashic field" or "A-field". He posits that the "quantum vacuum" is the fundamental energy and information-carrying field that informs not just the current universe, but all universes past and present (collectively, the "Metaverse").

László describes how such an informational field can explain why our universe appears to be fine-tuned as to form galaxies and conscious life forms; and why evolution is an informed, not random, process. He believes that the hypothesis solves several problems that emerge from quantum physics, especially non-locality and quantum entanglement.

We are used to thinking of our world as matter and energy. Our education systems teach physics, biology and chemistry to students who learn that energy is simply equal to mass through Einstein's famous equation: $E = mc^2$. Later on, some students might be lucky to learn something about information theory, information technology or communications, where information is treated as communication between two different systems, human or machine.

Medical and healthcare education is also based on similar "matter and energy" premises, so it is not surprising to find out that much of modern medicine and healthcare is based upon physical and chemical interventions. At the other extreme, information technologists take information for granted, as invisible bits and bytes that pass magically through their systems and can be made more or less useful depending upon the models they use to design and run their systems. They never stop for a minute to ask what that information actually is, and where it originates—it just "is".

Einstein is famously known to have received his ideas as images in dreams and thought experiments, and frequently stated that he did not think in symbols, but in images. Most creative people, going back to the ancient Greek philosophers, are on record that they are interacting

---

[11] Both in his 2004 book, "Science and the Akashic Field: An Integral Theory of Everything", and in his latest work; "The Self Actualizing Cosmos", Inner Traditions, 2014.

with a realm of images that can be reduced to symbols in language and numbers. Psychologists such as Carl Jung explored this realm as the "collective subconscious", and quantum physicists such as David Bohm called it the "implicate order".

A similar phenomenon is reported by scientists exploring paranormal phenomena such as psychic communication, remote viewing, spontaneous healing, human energy fields, near death experiences (all scientifically validated though still not well explained) and extra sensory perception, clairvoyance, telepathy, ghosts (still under investigation but growing evidence of their existence). In other words, a world of mainly images and sounds along with experiences that are often described as "out of body" appears to exist in parallel with our physical and energetic world.

One of the trickiest things about investigating such phenomena, is that many of the experiments appear to require people with open minds for them to work. Active sceptics who attempt to disprove such phenomena succeed in failing to elicit the phenomena, thus acting out a self-fulfilling prophecy which proves nothing. It seems that there is some kind of "tuning" going on between those who can access the frequencies that give access to this parallel world of information.

From a classic model of the universe that comprised bits of matter rotating around each other in empty space, we have come to a 21st century synthesis which describes the universe as an entangled, holographic, non-locally connecting in-formation field. What Einstein described as spacetime, is actually a "zero point field" or "quantum vacuum" described by quantum and string theory physicists, and it is this unified field that is "made of" information.

As Laszlo puts it:

"This is the "in-formation" that structures the physical world, the information we grasp as the laws of nature. Without information the energy-waves and patterns of the universe would be as random and unstructured as the behaviour of a computer without its software. But the universe is not random and unstructured; it's precisely "in-formed." Would it be any the less precisely informed, complex systems could not have emerged in it, and we would not be here to ask how this on first sight highly improbable development could have come about."

Science suggests that our brains can open to this unified information field (or information based "A-field", as referred to in several recent books)[1], and that this can happen in many ways, from dreams to daydreaming to imaginative thought experiments to spiritual experiences. Not only do our brains enable us to tune in to this latent information all around us, our bodies and all of nature are already doing so to synchronize their activities and stay alive. From our DNA synchronizing its activities through the emission of photons to our ability to sense the state of another human being or animal, we use this ability every moment without giving it a second thought.

What would it mean for the future of our world to start from an understanding of information as the basis for our universe, rather than matter and energy bumping into each other in "spacetime"? We would see things very differently, and certainly a lot more creatively and productively.

An information-based model of the universe based on the scientific breakthroughs of the past four decades creates the opportunity to transform the way in which we live, work, learn and develop. This model and its widespread implications are ripe for application in dozens of fields, through a variety of research and development programs. There is a great need for further research programs that can help provide the leadership and insights that make it possible for such research and development to be carried out in partnership with leading organizations and institutions globally.

To sum up, there are many promising areas for immediate application, including:

- **Biology, Medicine and Longevity**—Science is validating the information-based model of our bodies, health and healing as being highly effective, resulting in dramatic savings in drugs, hospitalisation and medical costs in general.[12][li] This is known by many names, including "MindBody Medicine". Our understanding of how our bodies work is also dramatically simplified if we apply the information-based model to cellular biology, as Dr Bruce Lipton[lii] and others such as Dr Elisabet Sahtouris have demonstrated

---

[12] Although there are charlatans in every field, including information healing and medicine, the discerning individual can find hundreds of decent scientific studies validating the effectiveness of this approach, and its superiority in many cases over traditional medical treatments.

with reference to the "Gaia theory" of Professor James Lovelock.[liii] There is a worldwide shift going on in which the healing paradigm is advancing from a mechanistic worldview to a living universe worldview, where the universe, our cells, organs and bodies are seen to be made of information and energy interacting to create the world we experience.

As a result of this information-based model, we find it also becomes possible to beneficially influence our bodies and energy systems through the state of our consciousness and through our intentionality. This new perspective then opens our eyes to the possibilities of energy and alternative medicines, where healing occurs at the cellular level. People can then learn to trust the innate intelligence in their bodies and cells, and heal through information rather than chemicals. This involves people becoming much more sensitive to the signals they receive from their bodies, and appreciating the different ways in which complementary and alternative medicine and therapies can help them heal faster and better, with fewer side effects.

Renowned neuro-surgeon Dr Norman Shealy's most recent discovery is that electrotherapy, can cause the telomeres, the tail of the DNA (the degeneration of which causes ageing), to regenerate itself. Telomeres ordinarily shrink by 1 percent each year. Shealy's technique suggests that electrotherapy can actually have them grow 4 percent a year.[liv] Living longer, healthier, more fruitful lives as a result of bio-information-based approaches to healthy living, medicine, longevity and healing is today a practical possibility for all people who take an interest in their own health and wellbeing.

- **Understanding Consciousness and the Mind**—The information-based model appears to give us a better explanation of how the mind works, and what can be done to promote the growth of vibrant lives and healthy minds. As Neuroscientist Dr Daniel Siegel points out: "[Mind is] an embodied and relational process that regulates the flow of energy and information, consciousness included. Mind is shared between people. It isn't something you own; we are profoundly interconnected. We need to make maps of We because We is what me is!"

As Dr Deepak Chopra points out, by taking this definition of mind one word at a time, we can explain the mind and consciousness in a single paragraph:

"The mind makes itself known through an organ of the body, the brain ("Embodied"). Our minds reflect the environment around us. We are constantly being shaped by the people around us, responding to their habits, speech, gestures and facial expressions ("Relational"). Mind is an activity. It isn't static but dynamic ("Process"). The jumble of data the universe produces would be chaotic unless something organised it into a coherent reality. To keep reality intact, each part must be regulated with every other part. ("Regulate"). There is an uninterrupted stream of consciousness to parallel the uninterrupted stream of external events. ("Flow"). To keep the flow going takes energy, at all levels from the immensity of the Big Bang to the micro level of ions passing through the cell membrane of a neuron. ("Energy"). Every piece of data can be seen as information, containing a bit of meaning. ("Information")."[lv]

The development of healthier, more creative minds and hearts, and more effective treatments for psychological illnesses, are both high potential applications of the information-based model. The fully integrated human being capable of thriving and contributing to the flourishing of those they interact with is increasingly a reality for millions of people who are benefitting from such insights into our mind and consciousness.

- **Paranormal Phenomena Demystified**—Scientists are now finding that there are ways in which the effects of microscopic entanglements "scale up" into our macroscopic world. Entangled connections between carefully prepared atomic-sized objects can persist over many miles. There are theoretical descriptions showing how tasks can be accomplished by entangled groups without the members of the group communicating with each other in any conventional way. Some scientists suggest that the remarkable degree of coherence displayed in living systems might depend in some fundamental way on quantum effects like entanglement. Others suggest that conscious awareness is caused or related in some important way to entangled particles in the brain. Some even propose that the entire universe is a single, self-entangled object.

What if these speculations are correct? What would human experience be like in such an interconnected universe? Would we occasionally have numinous feelings of connectedness with loved ones at a distance? Would such experiences evoke a feeling of awe that there's more to reality than common sense implies? Could "entangled minds" result in the experience of your hearing the telephone ring and somehow knowing—instantly—who's calling? If we did have such experiences, could they be due to real information that somehow bypassed the usual sensory channels?[lvi]

Scientists such as Dean Radin and parapsychologists have been researching these phenomena for decades, and some surprising evidence has emerged to explain paranormal phenomena. Parapsychology is the scientific study of three kinds of unusual events: Extra Sensory Perception, Mind-Matter Interaction, and Survival after Bodily Death. The existence of these phenomena suggest that the strict subjective/objective dichotomy proposed by the classic scientific paradigm may not be quite so clear-cut as once thought. Instead, these phenomena may be part of a spectrum of what is possible, with some events and experiences occasionally falling between purely subjective and purely objective.

Most parapsychologists today expect that further research will eventually explain these anomalies in scientific terms, although it is not clear whether they can be fully understood without significant (some might say revolutionary) expansions of the current state of scientific knowledge. There is a real opportunity to bring what are often regarded as fringe phenomena and experiences into the mainstream of psychology, validating those aspects of the phenomena susceptible to scientific investigation and improving our ability to understand and work productively with them. Organizations such as IONS ("Institute of Noetic Sciences") are leaders in this field of application.

- **Human Flourishing—Distributed Spirituality and Activism Catalyzed**—As the influence of traditional religion slowly wanes in the more developed parts of the world, religious beliefs have been abandoned without being replaced by new ways of creating meaning, caring and connection between people. Although the modernist and post-modernist worlds have replaced religious belief with the worship of science/progress, social/natural justice respectively, both of these appear to have created a large number of individuals who feel alienated, disconnected and dysfunctional. This is exemplified by the large number of cases of depression, anxiety and general psychological dis-ease experienced by North Americans, some

Europeans and their developing world counterparts in Asia-Pacific, the Middle-East and Africa. Clinicians have in effect diagnosed an existential crisis of epidemic proportions in these parts of the world. Psychologists have developed extensive programs focused on human flourishing and happiness, which emphasise the need for positive emotion, engagement, healthy relationships, meaning and accomplishment (known as the "PERMA" model)[lvii] Human thriving is itself largely an outcome of the processes of human consciousness, so anything that can shed more light on how our minds can produce thriving and resourceful states and intentions that lead to thrival, would be most welcome in comparison with much of the self-help and spiritual literature which remains anecdotal, sometimes confusing and often misleading.

In addition to some of the more tangible differences one can see emerging from an information-based model of the universe, would be a deeper understanding of peak experiences and altered states that would shed light on the potential of our species. While it is wise to remember that all such peak experiences and altered states will be interpreted by their experiencers through the lens of their current stage of development and worldview, (and are thus liable to various kinds of distortion), the impetus that such experiences give to human development is undeniable.

The development of the practical interconnections between the sciences of human flourishing and the next stages in the evolution of human spirituality is a further high potential area. Religion and spirituality are important conveyor belts for the evolution of human consciousness and capability, and they must evolve rapidly beyond their ethnocentric roots toward a global "transcend and include" perspective

- **Integral Models of Human Development**—as our species evolves toward a global civilization with the power of the gods to alter the face of our planet in the Anthropocene era, it is critical that we evolve a world-centric, integral consciousness that embodies integral ideals and values. "Aha" moments and peak states can be experienced at every one of a dozen different levels of development, and there is a responsibility accompanying any activity that involves generating peak states, to ensure that such states encourage the development of the individual and those in their world, rather than serve as an escape from the world.

While hundreds of techniques from meditation through to headphones with binaural beats are recommended by therapists, developmental coaches and a variety of teachers and gurus, most of these techniques are not embedded in scientifically validated developmental frameworks.

While integral philosophers[lviii] and developmental psychologists[lix] offer a range of developmental frameworks which concur around 8 to 10 developmental stages, Ken Wilber is one of the few who have mapped developmental practices to stages of development. Along with work by other pioneers in developmental research[lx], this work has resulted in effective practices such as ITP (Integral Transformative Practice) and ILP (Integral Life Practice).

Another important aspect of Integral Models of Human Development is the need for research and development activities to include personal, behavioural, cultural and socio-economic dimensions in order to effectively comprehend and intervene in complex, evolving human systems. At its root, integral philosophy and practice revolves around the art of shifting, balancing and integrating multiple perspectives.

The merging of the work of the greatest contemporary integral philosophers, Ervin Laszlo and Ken Wilber, into a coherent set of principles for lifestyles of health and sustainability, and a more just, civilized world, would dramatically catalyze the progress we need in these areas.

- **Education and Innovation**—Educating future generations using the information-based model would be much more appropriate for living in a world where knowledge and information are some of the key ways in which new forms of sustainable value can be created. Creating a post-materialistic, green growth experience economy requires a very different mindset and set of intellectual tools to one where agricultural and industrial processes are the predominant sources of wealth.

  The ability to blend both intuitive and rational processes together into a creative, innovative approach to one's life and work would seem to be a very high priority, certainly more important than smashing atoms together at the speed of light to prove that the "God-particle"

The Trouble with Paradise

(the infamous Higgs boson), exists. Yet for many in the mainstream the latter still appears to be the summit of intellectual achievement, though its practical consequences appear to be limited to justifying an out of date model of physics and the universe, and not much else at this point.

Some pioneering examples of educational institutions inspired by the information-based model have been created in recent years, but there is a very long way to go to shift the educational system from its modernist roots through its postmodernist cleverness to a form of integral wisdom that is desperately needed in a globalizing world.[lxi] Major challenges remain, however, in terms of both funding such initiatives and getting them into the mainstream of the educational sector, and these challenges must be overcome in order to ensure our educational systems evolve into the vehicles of human development and transformation they are capable of becoming.

New inclusive models of education for a globalized world that build on the core strengths of people and their communities are key to the evolution of worldcentric, thriveability consciousness. Such education models should empower learners to develop both as local civic minded culturally creative citizens and globally aware users of relevant enabling technologies and practices. Beneficial social and technological innovations are more likely to arise in such communities than in those where traditional educational "stovepipes" reinforce the industrial age paradigms.

- **Embedding ThriveAbility: Growing Green Economies**—How can we catalyze a thriving, sustainable future? In order for business to be a key engine of transformation to post-materialist, green growth economies, business leaders must change the game, change the rules and change their goals. The information-based model provides a starting point for such an initiative, together with the integral model of development which requires a balanced perspective to be applied to all human activities and business processes.

A new approach to business and organisations called "ThriveAbility"[13] has evolved to integrate strategy, innovation, sustainability and

---

[13] Originally conceived by myself in 2012, and developed further with my colleagues in the ThriveAbility Consortium. See www.embeddingthriveability.org. ThriveAbility is funded by a not-for-profit: www.thriveability.eu.

design in a way that simplifies the task and gets the job done while creating shared advantage and value. ThriveAbility is both descriptive and prescriptive, as well as predictive—it describes the emerging edge of leading practice in strategy, innovation, sustainability and design, while prescribing an approach which includes an equation, decision framework, process and dashboard for ThriveAbility.

ThriveAbility is predictive in the sense that it enables us to calculate the costs and benefits of different options and trade-offs in the design, development and scaling of new products, services and experiences. They key question is how to most effectively embed ThriveAbility into organisations, and alongside that, how to develop the methods, tools and information systems required to put this into practice.

Moving beyond sustainability with its focus on impact minimization, to ThriveAbility[14] with its focus on thrival maximization, is becoming an urgent imperative if we wish to ensure 9 billion people living well on earth in 2050. Research is required into how businesses and policymakers can radically simplify the task while accelerating investment into sustainable innovation, design and strategy.

Clearly how we think about who are as species, and where we are heading along our evolutionary trajectory, will again be redefined as this new generation of philosophers and the scientists mentioned in the next chapter will reveal. We have only begun to unravel the Ariadne's thread of the mystery of who we are, why we are here and where we are headed.

---

[14] To find out more about ThriveAbility please go to the Epilogue starting on page (291).

Chapter Eight
New Answers to Age Old Mysteries

We live in a mysterious universe. Although our species has made great strides in understanding the Kosmos since a famous Greek philosopher suggested that everything in our world is made of atoms a few thousand years ago, we are still only at the very beginning of our journey of fully understanding what really makes our universe, and us, tick.

Philosophers and sages throughout the ages have always speculated about the nature of our world and its place in the cosmos. The ancient Greek philosopher Plato is famous for his idea that our world is a "projection" of an ideal world in which all the perfect geometric forms exist. (Plato though everything in our world was ultimately made of triangles, as the Greeks were excellent at the mathematics of triangles). Plato's pupil Aristotle, however, preferred Democritus' idea that the world is made up of atoms, and that forms arise naturally in this world and develop accordingly. This classic tension between an "idealistic" world view and a "pragmatic" world view is still much in evidence today, although science has gradually enabled us to form a much more accurate picture of the world we live in and how it works.

A great many theories have been developed and experiments performed to test those theories, to establish some of the basic "laws" of physics, chemistry, biology and psychology, yet it is clear that the existing scientific approach to explaining our universe, materialism, has hit its limits.

In the past few decades, the unexplainable exceptions to the "laws" of these disciplines have increased dramatically in number and quality, while mainstream materialist science has been unable to offer a credible explanation for such anomalies. While hundreds of excellent books have been published detailing these anomalies and the "paradigm shift" taking place in science and allied disciplines, they are, for the most part, geared to specialist audiences who understand the highly technical terminology required to explain both the science and the anomalies.

In the pages that follow I hope to explain how these scientific breakthroughs help us to make sense of the world beyond our senses, and explain some of the mysteries of the kosmos in plain English.

The Classic Scientific Method—
Establishing Propositional Truths

Despite centuries of scientific progress, the central idea of the scientific method has remained shrouded in mystery for the layman. Even school children studying science do not seem to be given an adequate explanation as to what the scientific method is, how it works and why it is of paramount importance in our lives.

This is a great pity, as just about everything we rely on in modern civilization was developed as a result of the use of the scientific method, whether in a high tech laboratory or in a workshop somewhere. We are all familiar with the idea of a "test"—we learn something, and then we are tested on that knowledge. Depending upon how well we have mastered our topic or project, we pass or fail the test.

The same applies to any new knowledge or practice—at first, we notice something interesting or useful about the natural or social world, and we describe this new phenomenon by advancing a theory about what is causing it. Lightning, for example, is considered by superstitious tribes people to be the wrath of the gods, and in Africa "witches" are sought out and burned after a village is hit by lightning because they are believed to be causing the lightning through "muti" or black magic.

Here is the simple theory used by such tribes: "Gods and witches acting as their agent = lightning" and their "solution" to this problem is to burn the witch/es = "No more lightning".

Thanks to many scientists who have risked their lives to study lightning over the centuries, we now know that lightning is a plasma which results from an electric current passing between a negatively charged cloud and the positively charged ground, superheating the air causing very bright light (lightning) and noise (thunder). We invent lightning conductors as a solution, and the problem is "solved".

That might all appear rather obvious to anyone who has completed high school, but the science and technology required to create both the theory and the solution are only very recent in the history of our species. Many thousands of people using the scientific method to test different theories about lightning were required to go from the "witches causing the gods to strike our village with lightning" explanation, to "lightning is a plasma which results from an electric current passing between a negatively charged cloud and the positively charged ground, superheating the air causing very bright light (lightning) and noise (thunder)".

Indeed, we had to first develop a science of weather, whereby the dynamics of thunderclouds could be properly understood. Next we had to develop a science of electromagnetism to explain why a bolt of lightning appears to jump from a cloud to the earth (though in reality the first "strike" often starts in the ground and goes toward the cloud.) And finally, we needed to develop sophisticated instruments to measure the charge, temperature and speed of lightning to understand that it is a plasma, which is in itself a fourth form of matter and energy beyond solids, liquids and gases. Fire and our sun itself are also plasmas, and were themselves mysteries that went unexplained for many millennia before modern science arose.

The history of our species shows that we worship what we desire and fear, and do not understand. Paleolithic tribes began to worship nature gods somewhere between fifty to one hundred thousand years ago, marking the beginning of the religious mind. We painted cave walls with images of the animals we needed to hunt for our survival, then later on we built primitive temples in which we sacrificed animals and humans to the gods to ensure the rains came and the harvests were good.

During the agricultural revolution some ten to five thousand years ago, we began to systematically breed crops and animals that were more effective in meeting our needs, enabling us to build up surpluses and cities—the first "civilizations". With the emergence of writing, we were able to record what worked, and become even more systematic in our attempts to copy what worked, and abandon those practices that did not.

Yet even during the Axial age some three thousand years ago, the early Chinese, Babylonian, Egyptian and Greek philosophers were still struggling to understand how our world works. From Plato to NATO, we have also devoted a considerable proportion of our efforts to defending ourselves from invaders, and to attacking and plundering other civilizations for their surplus food, women and "stuff". In fact, the need to develop effective weaponry was just as important as the need to domesticate animals and develop technologies to work the soil and hunt more effectively.

The scientific method comes in many forms, but at its most basic it is a process of experimentation, conscious or unconscious. From experimenting with different ways in which to fragment stone to make sharp Acheulian hand axes, to trying different breeding strategies for animals and plants to get healthier and more abundant progeny, we have always been experimenting since our large brains endowed us with imagination hundreds of thousands of years ago.

This "What-if?" capacity lies at the heart of all progress and innovation, but it has taken many centuries for human kind to learn how to systematically harness the raw power of our imaginative faculties. And our imaginative capacities are only as good as the information and knowledge they work with—as the saying goes: "Garbage in, garbage out". Our imaginations are just as good as producing elaborate fantasies divorced from reality and which become counter-productive and sometimes deadly, whether it is the dream of a Third Reich or a the dream of "true believers" of one religious creed or the other of a world of converts to their fundamentalist beliefs. "I have a dream" is not always a good thing.

Since the dawn of the emergence of human consciousness, we have had the capacity to ask questions to which there were few, if any, readily available answers. The sun rises in the East and sets in the West; the stars appear at night; animals are born and die, and some migrate; warm summers are replaced by cold winters; and we humans are all born, and we all die, sometimes naturally and often due to mysterious diseases. These basic facts have not changed at all since the dawn of human consciousness, although our explanations for them have evolved from primitive myths to scientific fields of enquiry in which many good answers and useful technologies are now available.

The scientific method is traditionally associated with our ability to establish truths about the material world through a process of observation about a particular phenomenon, theory formulation, theory testing ("experiments") and validation or negation of the theory based upon the success or failure of the experiment to validate the specific theory about the phenomenon being observed. Here follows a brief recap of the past few millennia of scientific and technological progress, a short list of what we have already learned in key fields of science, thanks to the scientific method in its classic sense of experimentation, together with a healthy dose of serendipity.

Twenty World-Changing Discoveries- Learning About Our Exteriors

Listed in the order they were discovered, here are twenty examples of major scientific and technological breakthroughs of our species and their consequences. It is interesting to note that apart from the discovery of fire several hundred thousand years ago, most of these discoveries and inventions have occurred since the end of the last Ice Age ten thousand years ago.

Fire

We don't know the identity of the experimenter or experimenters in the Acheulian culture in Africa who discovered how to start, control and utilize fire about 790,000 years ago. But their mastery of rapid oxidation was one of the most important developments that sustained the survival and spread of humanity.

The invention equipped early humans with a scary deterrent—flaming torches—to protect them and their vulnerable young from predators. It also provided a source of warmth that helped them to survive temperature downturns. Additionally, the ability to cook animal flesh and vegetation increased food choices for humans and helped them to avoid malnutrition. Perhaps more than any other invention, fire was the breakthrough that enabled humans to multiply and spread across the planet's surface.

Agriculture

Agriculture really is not one, but a series, of scientific and technical breakthroughs—such as the development of irrigation technologies, and the invention of crop rotation and fertilizers—that occurred over thousands of years. But it all started when humans figured out how to gather seeds from wild plants, plant and tend them, and harvest them. According to DNA analysis of modern foodstuffs, development of the "founder crops"—wheat, barley, chickpeas, lentils, flax and others—dates back about 9,000 to 10,000 years in southwest Asia.

The Wheel

The earliest actual evidence of a wheel in human history occurs at about 3500 B.C. in Mesopotamia. However, this evidence is associated with the wheel's use in pottery-making, not as a tool for transportation. It took another 300 years or so for the people of Mesopotamia to realize that the wheel could also help them to move things from place to place.

Wheels evolved in a few stages throughout history, beginning with the use of logs as rollers to facilitate transportation and continuing on through the replacement of rollers with wheels that rotate on an axle. By 2000 B.C., wheeled chariots appear in the archaeological record throughout ancient Egypt. Only by then the wheels had spokes, making them considerably stronger and lighter.

The wheel was probably the most important mechanical invention of all time. Just about all modern mechanical devices use the wheel in some way—cars, buses, bicycles, factory machines, toys, wristwatches, aircraft, generators and more.

The Copernican System

In 1543, while on his deathbed, Polish astronomer Nicholas Copernicus published his theory that the Sun is a motionless body at the center of the solar system, with the planets revolving around it. Before the Copernican system was introduced, astronomers believed the Earth was at the center of the universe. The publication of Copernicus' book, "On the Revolutions of the Celestial Spheres", is considered a major event in the history of science. It began the Copernican Revolution and contributed importantly to the scientific revolution that followed and challenged the supremacy of the Church.

It took half a century for the work of Kepler and Galileo to provide substantial evidence defending Copernicanism. It was not until after Isaac Newton formulated the universal law of gravitation and the laws of mechanics which unified terrestrial and celestial mechanics, that the heliocentric view became generally accepted.

Gravity

Isaac Newton, an English mathematician and physicist, is considered the greatest scientist of all time. Among his many discoveries, the most important is probably his law of universal gravitation. In 1664, Newton figured out that gravity is the force that draws objects toward each other. It explained why things fall down to earth and why the planets orbit around the Sun.

Vaccines

Of all the scientific breakthroughs that have transformed humanity, the development of vaccines is undoubtedly among the most important in terms of human survival. Diphtheria, polio, rubella and whooping cough are just a few of the diseases that can be prevented with vaccines. In fact, some diseases have been completely eradicated through the use of vaccines. One example is smallpox, which once killed 35 percent of its victims [source: History of Vaccines].

In 1796, English physician Edward Jenner determined that injecting a boy with pus from cowpox blisters (a mild disease transmitted by cows) prevented him from contracting the much deadlier smallpox. At first Jenner's discovery attracted doubt and ridicule but later he was proved to be right and vaccinations became widespread. There is no doubt that humans are living longer and healthier lives thanks to the development of vaccines.

Electricity

Michael Faraday made two big discoveries that helped create the modern world. In 1821, he discovered that when a wire carrying an electric current is placed next to a single magnetic pole, the wire will rotate. This led to the development of the electric motor. Ten years later, he became the first person to produce an electric current by moving a wire through a magnetic field. Faraday's experiment created the first generator, the forerunner of the huge generators that produce our electricity.

Trying to imagine what life was like before electricity is pretty difficult. Electricity is used to manufacture our homes and cars and it powers everything from appliances to streetlights. Before the discovery and development of electricity, not only did we do nearly every task and household chore by hand, we pretty much had to do them before nightfall. Electricity is also responsible for making our modern high-tech world possible. It charges our phones, computers, and virtually every other tool of communication used in modern life.

Evolution

When Charles Darwin, the British naturalist, came up with the theory of evolution in 1859, he changed our idea of how life on earth developed. Darwin argued that all organisms evolve, or change, very slowly over time. These changes are adaptations that allow a species to survive in its environment. These adaptations happen by chance. If a species doesn't adapt, it may become extinct. He called this process natural selection, but it is often called (incorrectly), the "survival of the fittest".

The inheritance of acquired characteristics (tallness, for example), turns out to be possible across several generations without having to resort to genetic or neo-Darwinian explanations (in fact Darwin and his grandfather Erasmus subscribed to this theory). Darwin's views about human evolution have also been consistently misrepresented. Politicians and economists seized on Darwin's "Origin of the Species" to justify their idea that we are rational, utility optimizing beings who evaluate trade-offs between goods and options, and choose the one that best serves our narrow, selfish interests. During the 1980's it became popular for politicians from Margaret Thatcher to Ronald Reagan to praise such individualism and enterprise as the backbone of the global economy.

Again, very few who cite Darwin know that his final work, entitled: "The Descent of Man", argued that evolutionary love and higher morality were, for Darwin, the drivers of humanity's future evolution, a thought echoed by Barbara Marx-Hubbard a century later in her philosophy of Conscious Evolution.

The Germ Theory of Disease

Before French chemist Louis Pasteur began experimenting with bacteria in the 1860s, people did not know what caused disease. He not only discovered that disease came from microorganisms, but he also realized that bacteria could be killed by heat and disinfectant. This idea caused doctors to wash their hands and sterilize their instruments, which has saved millions of lives.

Plastic

Plastic took a big leap forward in the early 20th century with the invention of Bakelite, the first material to be made entirely by man. The inventor, Belgian chemist Leo Hendrik Baekeland, mixed phenol and formaldehyde together and subjected them to heat and pressure. Seemingly overnight, this material was used in electronics, kitchen appliances, jewelry, toys, and countless other products.

Today, plastic is incorporated into so much of daily life—everything from toothbrushes to satellites and the computer you're using to read this article—that it's difficult to imagine a world without it. However, plastic does have a dark side. Plastics are implicated in a number of environment disasters, including giant flotillas of plastic debris in the oceans and harmful pollutants such as BPA in our bodies. Nevertheless, plastic is such a part of life that ways are being found to curtail its harmful effects rather than eliminate it entirely.

Theory of Relativity

Albert Einstein's theory of special relativity, which he published in 1905, explains the relationships between speed, time and distance. The complicated theory states that the speed of light always remains the same—186,000 miles/second (300,000 km/second) regardless of how fast someone or something is moving toward or away from it. This theory became the foundation for much of modern science, although to date scientists have failed to reconcile the standard model of physics that emerged from the general theory of relativity, and the quantum theory that describes life at the micro-scales.

General relativity applies to the large scale structures of the universe—planets, stars and galaxies—and describes gravitation, one of the four "fundamental forces". Quantum mechanics describes the other three forces (electromagnetism, and the strong and weak nuclear forces). Recent attempts to reconcile the theories of relativity and the quantum such as superstring and M-theories require ten and eleven dimensions to the universe, as well as 10^{500} different kinds of universes. This runaway proliferation of unobservable universes (except our own) is not only untestable, but also seems improbable, failing to unify anything.

The Big Bang Theory

Nobody knows exactly how the universe came into existence, but many scientists believe that it happened about 13.7 billion years ago with a massive explosion, called the Big Bang. In 1927, Georges Lemaître proposed the Big Bang theory of the universe. The theory says that all the matter in the universe was originally compressed into a tiny point the size of an electron, known as a "singularity". After its initial appearance, it apparently inflated (the "Big Bang"), expanded and cooled, going from very, very small and very, very hot, to the size and temperature of our current universe. This event marked the beginning of both space and time as we know them.

George Smoot and his colleagues were awarded the Nobel Prize in 2006 for their observations of the background cosmic radiation which confirms the Big Bang theory.

Penicillin and Antibiotics

For most of human history, virtually everyone on the planet faced the risk of dying in epidemics of bacterial diseases that sometimes ravaged multiple continents. One such disease, Bubonic plague—the "Black Death"—killed an estimated 200 million people in the 14th century alone

Antibiotics are powerful drugs that kill dangerous bacteria in our bodies that cause illness. In 1928, Alexander Fleming discovered the first antibiotic, penicillin, which he grew in his lab using mold and fungi. Without antibiotics, infections like strep throat could be deadly. Since then, the use of penicillin and other antibiotics has led to dramatic reductions in the death rate from certain once-common diseases like syphilis, gangrene, scarlet fever, gonorrhea, and tuberculosis.

DNA

The discovery of DNA—the very building blocks of all life—is arguably one of the most important discoveries of the 20th century. DNA, the acronym for deoxyribonucleic acid, was first discovery in the late 1860s by a Swiss chemist named Friedrich Miescher, though much of our understanding of DNA comes from the work of James Watson and Francis Crick, who explained the double helix structure of a DNA strand in 1953.

DNA is made up of two strands that twist around each other and have an almost endless variety of chemical patterns that create instructions to make proteins for the human body. Our genes are made of DNA and determine things like what color hair and eyes we'll have. Watson and Crick won the Nobel Prize for their work. The discovery has helped doctors understand diseases and may someday help prevent some illnesses like heart disease and cancer, although work in epigenetics now seems more promising.

Periodic Table

The Periodic Table is based on the 1869 Periodic Law proposed by Russian chemist Dmitry Mendeleev. He had noticed that, when arranged by atomic weight, the chemical elements lined up to form groups with similar properties. He was able to use this to predict the existence of undiscovered elements and note errors in atomic weights. In 1913, Henry Moseley of England confirmed that the table could be made more accurate by arranging the elements by atomic number, which is the number of protons in an atom of the element.

X-Rays

Wilhelm Roentgen, a German physicist, discovered X-rays in 1895. X-rays go right through some substances, like flesh and wood, but are stopped by others, such as bones and lead. This allows them to be used to see broken bones or explosives inside suitcases, which makes them useful for doctors and security officers. For this discovery, Roentgen was awarded the first-ever Nobel Prize in Physics in 1901.

Quantum Theory

Danish physicist Niels Bohr is considered one of the most important figures in modern physics. He won a 1922 Nobel Prize in Physics for his research on the structure of an atom and for his work in the development of the quantum theory. Although he help develop the atomic bomb, he frequently promoted the use of atomic power for peaceful purposes.

Atomic Bomb and Nuclear Power

The legacy of the atomic bomb is mixed: it successfully put an end to World War II, but ushered in the nuclear arms race. Some of the greatest scientists of the time gathered in the early 1940s to figure out how to refine uranium and build an atomic bomb. Their work was called the Manhattan Project. In 1945, the U.S. dropped atomic bombs on the Japanese cities of Hiroshima and Nagasaki. Tens of thousands of civilians were instantly killed, and Japan surrendered. These remain the only two nuclear bombs ever used in battle. Several of the scientists who worked on the Manhattan Project later urged the government to use nuclear power for peaceful purposes only. Nevertheless, many countries continue to stockpile nuclear weapons. Some people say the massive devastation that could result from nuclear weapons actually prevents countries from using them.

Information Technology

Humans have been storing, retrieving, manipulating and communicating information since the Sumerians in Mesopotamia developed writing in about 3000 BC, but the term "information technology" in its modern sense first appeared in a 1958 article published in the Harvard Business Review; authors Leavitt and Whisler commented that "the new technology does not yet have a single established name. We shall call it information technology (IT)." Based on the storage and processing technologies employed, it is possible to distinguish four distinct phases of IT development: pre-mechanical (3000 BC-1450 AD), mechanical (1450-1840), electromechanical (1840-1940) and electronic (1940-present).

The first electronic digital computers were developed between 1940 and 1945, originally the size of a large room, consuming as much power as several hundred modern personal computers (PCs). In this era mechanical analog computers were used for military applications.

Modern computers based on integrated circuits are millions to billions of times more capable than the early machines, and occupy a fraction of the space. Simple computers are small enough to fit into mobile devices, and mobile computers can be powered by small batteries. Personal computers in their various forms are icons of the Information Age and are what most people think of as "computers." However, the embedded computers found in many devices from MP3 players to fighter aircraft and from toys to industrial robots are the most numerous.

In a world where half of the people on earth have a mobile phone, and a third have internet access, it is hard to imagine a time when microchips did not exist, yet they have only been in use for just over four decades. IT will continue to transform our world for centuries to come.

Higgs Boson

In 2012, physicists at the Large Hadron Collider ended a 5-decade-long search when they announced the discovery of the Higgs boson. This long-sought particle is responsible for giving all other subatomic elements, such as protons and electrons, their mass, and was the final piece in the Standard Model, which describes the interactions of all known particles and forces. While LHC researchers were cautious, only calling their results a "Higgs-like" particle until more data and analysis is available, the finding was widely hailed as the most important fundamental physics discovery in more than a generation.

Scientists had been hoping that spotting the Higgs would also provide their first glimpse of physics beyond the Standard Model, which has various problems and inconsistencies that need fixing. But the particle has so far proved to be stubbornly normal, with little to no deviation from what was predicted under the Standard Model. It may be that physics has reached something of a dead-end with the atom smashing approach with ever costlier accelerators, and that we need a new way of understanding our universe that yields real predictive power and also explains many things which are today still regarded as "mysteries".

Expanding the Scientific Method Beyond the World of Exteriors

These 20 major scientific discoveries, along with millions of other more minor ones, have helped one half of humanity master the "basics" of a comfortable, secure existence. No question about that! The bigger question now, however, is how we co-create a thriving future for all of us that doesn't cost the earth. The answers to that massive challenge lie in the new approaches we must develop to living sustainably and well. While each of these "exterior" scientific discoveries have helped us master the basics of a more secure, comfortable and disease free world, such an approach only covers the first few levels of Maslow's hierarchy of needs: security, food, shelter and respect, the first few rungs on the ladder of evolution for our species. These "basics" are outer needs, which do not address our major inner needs such as belonging, meaning, self-expression and development, fulfilment, self-actualisation and inner peace.

The great news about the evolution of science that is fuelling the second Renaissance, is that it represents a much more balanced approach to our inner and outer needs, with a real potential to catalyse the self-actualization of every human being on this planet. We are learning to apply the scientific method to our inner worlds just as effectively as we have applied it to our outer material selves and needs.

Yet there appears to be a major schism between materialists who reject any approach to exploring our interiors as "mysticism", and those of us who believe that our balanced self-development is one of the keys to a thriving future. Integral philosophers and intellectuals stress that we need to explore our "inner" worlds to achieve a leap in the evolution of our species. A big part of the problem is that this debate is framed in "either/or", "black/white" terms, rather than an "And/Transcend" approach to our inner and outer lives.

The good news is that this polarization is not inevitable—in fact, the latest science simply transcends this fruitless debate entirely, and that is what makes it so exciting and full of potential. Truth claims are not only reserved for the material, exterior "world out there"—post-modernist philosophers and scientists distinguish between three main versions of the scientific method:

- *Empirical methods of enquiry* based on the five senses and our extensions thereof (from microscopes to satellites), to establish truths about the material world of exteriors, such as physics, chemistry, biology and so on. Here we establish **propositional truths** about an objective, external state of affairs, classically known as objective **truth**;
- *Mental-phenomenological methods of enquiry* based on interior systems of meaning-making such as linguistics, mathematics, hermeneutics, logic and so on. Here we establish **cultural truths** about predictable truths in cultural contexts and spaces, classically known as **goodness**;
- *Transcendental methods of enquiry* based on traditions and practices such as contemplation, meditation, yoga and so on, whereby maps and models of the interior worlds and replicable processes for re-producing inner experiences are deduced and tested. Here we establish **interior or transcendental truths**, based upon the reliability of the interpretations, maps and models established by instructing and testing subjective states, classically perceived as **beauty**.

In our daily lives we encounter and use each of these kinds of truth frequently—for example, in going on a holiday we may rely on objective safety truth claims by airlines to ensure we arrive at our destination unharmed; we probably select our travel destination based on our inner perceptions of the goodness of the places and people at our destination (backed up by the perceptions of others on websites such as "Trip Advisor"); and we expect to be able to chill out and relax in the spa at our destination based on the predictable effects of therapeutic practices rooted in yoga, alternative medicine, meditation or simply closing one's eyes and taking several deep breaths of fresh air. A holiday is, in essence, a search for beauty and goodness at the right price to bring about a highly valued inner state we call relaxation or peace of mind.

Dr Robin Lincoln Wood

What Are the Big Questions for the 21st Century?

Who are the 21st century "paradigm-busters" blazing a path to a more coherent and useful explanation of what our world is, how it works and what becomes possible through this new understanding?

Despite the incredible progress we have made in our explorations as a species, we still struggle to answer some very basic questions with any degree of consensus, despite a great deal of evidence that the currently dominant, materialistic scientific paradigm is deeply flawed and overdue for a reboot. For example, biologist Rupert Sheldrake takes on the materialists in his latest book: "Science Set Free". He asks materialists ten hard questions, and then offers a great deal of compelling evidence to suggest that the "scientism" of the materialists is unable to answer these questions with a resounding "Yes!" or even a tentative "Perhaps".

Using compelling recent evidence and scientific discoveries, Sheldrake questions some fundamental assumptions of the materialistic paradigm, including:

- **Physics and Chemistry**—Is Nature Mechanical? Is the Total Amount of Matter and Energy Always the Same? Are the Laws of Nature Fixed?
- **Biology**—Is Matter Unconscious? Is Nature Purposeless? Is All Biological Inheritance Material?
- **Psychology**—Are Memories Stored as Material Traces? Are Minds Confined to Brains? Are Psychic Phenomena Illusory?
- **Medicine**—Is Mechanistic Medicine the Only Kind that Really Works?

Tens of thousands of impeccably conducted, predominantly double-blind randomized trial experiments over the past fifty years have demonstrated that there are fundamental flaws in the current dominant scientific paradigm, as Sheldrake demonstrates in "Science Set Free".

Yet the new scientific paradigm which has been emerging in the wake of the "weirdness" of quantum physics and the inability of the Multiverse and String Theories to agree on some fundamental aspects of how our universe arose, how it works and where and how it will end (if it ends at all), is still in the process of formation, with much scope for disagreement and further refinement.

We are used to thinking of our world as matter and energy. Our education systems teach physics, biology and chemistry to students who learn that energy is simply equal to mass through Einstein's famous equation: $E = mc^2$. Later on, some students might be lucky to learn something about information theory, information technology or communications, where information is treated as communication between two different systems, human or machine.

Medical and healthcare education is also based on similar "matter and energy" premises, so it is not surprising to find out that much of modern medicine and healthcare is based upon physical and chemical interventions. At the other extreme, information technologists take information for granted, as invisible bits and bytes that pass magically through their systems and can be made more or less useful depending upon the models they use to design and run their systems. They never stop for a minute to ask what that information actually is, and where it originates—it just "is".

What would it mean for the future of our world to start from an understanding of information as the basis for our universe, rather than matter and energy bumping into each other in "spacetime"? We would see things very differently, and certainly a lot more creatively and productively. Here is a short list of just some of the major topics of research currently being pursued by scientists around the world in this vein:

PHYSICS AND COSMOLOGY

- How does memory arise in and how is it stored in quantum and other fields?
- What is the physical basis of entanglement? What are the implications of spatial and temporal entanglement for physics?
- Is there coherence in galactic evolution?
- What evidence is there for theories of universe-cycles in the multiverse?
- Is there information transfer among universe-cycles?
- What proof is there for concepts of a "hidden" dimension in this universe or in parallel universes?

LIFE AND SOCIAL SCIENCES

- What are the roots of so-called "inborn behavior" and "instinct"?
- How does coherence arise in the transformation of genetic information in biological evolution?
- How can we explain coherent or parallel evolution in distant and non-communicating cultures and societies?
- How does complementary medicine deliver spontaneous healing?

CONSCIOUSNESS RESEARCH

- Where does the information we can access in meditative or altered states come from?
- How do remote viewing, telepathy, and related psychic phenomena work?
- How can we best explain NDEs (near-death experiences), OBEs (out of body experiences) and other phenomena of discarnate consciousness?
- What are the dynamics of human consciousness and what is "Super" consciousness?

METAPHYSICS

- What are the key synchronicities/convergences between modern science & traditional wisdom, and what are their implications?
- Is there direction and purpose in the unfolding of universe-cycles?
- Is it possible to experience the presence of cosmic or divine intelligence—and what exactly is such an intelligence?
- How can we more effectively explore our intuitions and concepts of soul or spirit?

Chapter Nine
Overview of a Transpersonal Learning Journey

These are all important and fascinating questions, and no doubt researchers will eventually succeed in answering many or even most of them, if we can make it into the 22nd century. Each of them deserve a separate book to even begin to get to grips with a detailed explanation and understanding of the phenomena involved, and we are now close to the end of this one.

I promised in the Introduction to this book that I would make what can often be very dry topics a little more fun by injecting some titbits of biographical information and stories into the text, and I have done so sparingly in the interests of keeping this book to a manageable length. It is in this spirit that I offer you the final instalment of a quasi—autobiographical update on how most of the topics we've been exploring have been played out in real life, as opposed to simply being words on a page, and just how useful they can be in appreciating the richness and magnificence of ourselves and our world.

Memory can also be a perfidious thing, as Kurt Vonnegut put it in "Slaughterhouse-Five":

"And Lot's wife, of course, was told not to look back where all those people and their homes had been. But she did look back, and I love her for that, because it was so human. So she was turned into a pillar of salt. So it goes."

So, my apologies to anyone who is disturbed, annoyed or even bored by this brief reconstruction of an (at least) eventful life—may I not be turned into a pillar of salt in turn for looking back to look forward.

I was born, kicking and screaming, in the bustling Indian Ocean port of Durban, South Africa, to two happy young parents eleven years after the end of the second World War. From all accounts, my charming, Reactive "Beige" self, fully equipped with a vigorous R-complex, was particularly good at expressing itself at all hours. My father was a rising young marketing executive at Lever Brothers, my mother the secretary to the Managing Director of the same bustling multinational enterprise, purveyor of fine soaps and skin care products to the house wives of Southern Africa and the world.

In 1956 South Africa was experiencing a surge of economic growth. Part of the greater British Commonwealth, it was the Age of Abundance and Growth, seemingly everywhere, with two billion people living comfortably within the limits of the planet. The future was bright, in fact the future was Achievist "Orange", built on a solid base of good old fashioned British Empire civil society Conformist "Blue" values, with a rising dash of Afrikaner (descendants of the original Dutch settlers) influence in politics and commerce. And this, of course, was Africa, with its teeming tribes of Zulus and Xhosas, Shangaans and Sothos, Vendas and Tswanas—26 tribes in all in South Africa alone. Together with the Indians, Pakistanis, Malaysians and many others who had landed on those shores courtesy of the Dutch, British and many other ships that plied the harbours of the Indian Ocean.

That same year, somewhere in Johannesburg, Nelson Mandela was practicing law with Oliver Tambo, also secretary of the African National Congress ("ANC"), but for a number of years I was blissfully unaware that the world I was born into had its own share of escalating major problems which would lead to my own exile as an adult in 1986.

I was apparently taken often to the beach as an infant, and developed an early love of water and the seaside. Possessed of a fearless and adventurous nature, I would later get myself into all sorts of minor troubles. Eighteen months later my younger brother arrived, then a younger sister. We were an active young family unit, and my Naïve "Purple" pride in my siblings was evident when I pushed them in their pram gently down the stairs (one step at a time), with my mother managing to catch us before any damage was done.

My first memories are of the house we moved into in Robindale, Johannesburg[15], and playing "Daddy and Mommy and Priest" with my "girlfriend" Vivienne and her brother Billy, who actually ended up becoming a priest himself. My insatiable curiosity meant that I became an early deconstructionist—I would take apart anything that moved, or

---

[15] Our house was in Friar Tuck Road, and because my name is Robin, my delight at living in a neighbourhood named after me knew no bounds.

made a noise, if left to my own devices. Radios, clocks, toy trains—my small fingers could get into just about anything.

Looking back now, it is clear that despite the occasional childhood illness and nephritis (a serious kidney disease from which I luckily recovered at age 4), I was born into paradise. We had a large garden with a small pool, which I repeatedly attempted to get into despite a wire fence, and I was taught to swim at the end of a broom stick in order to avoid what must have seemed to my parents to be my inevitable drowning. Despite numerous injuries climbing trees, knocking my front teeth out in a fall, piercing my cheek falling off parallel bars, and knocking myself out cold a few times, I made it through my first few years in one piece.

Then came the fatal day I was enrolled in nursery school. At "Little by Little", it became clear that my wild, Huck Finn ways were going to be a problem. Sitting on little chairs at tiny tables "behaving" had very little appeal for me, and I would refuse to come in off the swings after break, going higher and higher much to the consternation of Mrs Little and the staff. My young, Egocentric "Red" rebel had asserted itself, and I experienced my first dressing down in public. Man, was I in trouble.

The lovely people at Little by Little survived intact despite my shenanigans, and after a year there I started my glorious career in Grade 1, with an elderly spinster named "Mrs Priest" as my teacher. As strict as her name suggested, I was rapped on the knuckles with her ruler more than I care to remember. My distaste for neat letters and sitting up straight in my chair led me into a number of confrontations with authority, but I do remember the magic of numbers appealing to me greatly. I also developed an impressive array of animal impressions after several visits to the Zoo, which assured me of a devoted gang in the playground, even though this did not endear me to the Powers that Be.

Looking back on it, I was lucky to be the beneficiary of a secure, happy early childhood. My father had been one of the early supporters of the "Progressive Party", which demanded an end to apartheid. Yet trouble was brewing in South Africa, as the ruling National Party, the architects of "Apartheid", had to begin suppressing riots around the country against the discriminatory Apartheid system. Led by Mandela and many others in the ANC, such peaceful demonstrations sadly ended up turning into angry confrontations with heavily armed police, and after the deaths incurred in the Sharpeville riots, our family emigrated to Canada.

We left what was then Jan Smuts airport for London on a Lockheed Tristar in spring of 1963. By now, I had learned to be a "good boy", and was rewarded by the pilot letting me sit on his lap and "fly" the airplane. We stopped in Entebbe in Uganda where I saw my first crocodiles on Lake Victoria, then Nairobi, Cairo (a donkey ride around the pyramids) and Athens, where my brother and I raced up and down the steps of the Acropolis and sniggered at the large pom poms worn on the shoes of the Greek palace guards.

After further stops in Paris and Luxembourg, we ended up on a ferry to London. My father had always promised us that he would take us to have tea with the Queen when we arrived, and true to his word we ended up on a train to Windsor Castle after discovering that she was not in residence at Buckingham Palace. Sadly, it turned out that her busy schedule did not allow for our tea party after all.

The sheer grandeur of the sights I had seen had a great effect on my young mind, and I went from being "Jungle Boy" worshipping Tarzan, to a proud young supporter of the British Empire, gaining my first Conformist "Blue" stripes. My limbic system was getting a good daily workout, while my neocortex benefitted from an early dose of solidly liberal Canadian primary school education. And I learned to play ice hockey, badly.

Apart from several brief visits to Hell due to illnesses and accidents, my life had been heavily biased towards the Paradise end of the scales until the local Canadian bullies discovered me as an unwilling victim. At the same time our teacher began playing Beatles records to us in music class, I began to learn the rudiments of self-defence. At the age of eight I discovered real fear for the first time, and it was Hell.

A few things stand out from this brief pre-adolescent phase worth noting here. The first concerns the experience of being an "outsider", which had never been a problem before. The second concerns the nature of the relationship between different developmental lines and levels, and the need for a balance between them for healthy human development. And the third relates to the emergence of a felt sense of spirituality.

Being an outsider is a common theme in our human condition, though we do not all get to experience the unpleasant side effects of being bullied or hurt as a result. Pre-adolescents begin to form "gangs" of one kind or another, and as an outsider, you do not come pre-equipped with gang to protect you if things get a little rough in the playground.

As brilliant and progressive as the Canadian education system in Toronto was in the 1960's, the implications of things like emotional intelligence had yet to be discovered. The almighty "Intelligence Quotient" held firm sway over the halls of Academia, and I had been born into the equivalent of the Royal IQ family. In some strange quirk of fate, the first born son on my father's side of the family going to back to my great grandfather, had always been possessed of a genius IQ.

There is an amusing tale of my grandfather, Lincoln James, a patriotic gentleman, who signed up to "fight Hitler" in the South African Army in the early 1940's. During his officer training, the first IQ tests were being administered to the new recruits. One evening while sitting in his bungalow with the other recruits, Lincoln was summoned to the officer's mess, and made to stand at attention while the officers dined. After a respectful period, Lincoln asked exactly why he was needed. The commanding officer explained: "Wood, you've just scored higher than General Smuts (at that time the man with the highest IQ in South Africa) in the IQ tests, so we just wanted to take a look at you. Dismissed!"

Not surprisingly, our family developed something of an obsession with psychological tests, and psychology in general. This caught up with me when I was accelerated from Grade 2 to Grade 4 at my school, becoming two years younger on average than the rest of my classmates. While this was academically a good move, meaning I was less bored in school than I had been before and began to feel challenged, I immediately became a target for the bigger slower kids in school, for whom my coming first in class also meant I came first on the list of victims for the bullies. Sadly I had not yet developed an effective stand-up comedy routine such as Woody Allen had used to fend off his attackers at school.

Today we would measure the emotional maturity of a child before consigning them to this fate, but EQ did not then exist as a concept. Years later the teachers admitted that acceleration had been a mistake, but I learned to distract and eventually avoid the bullies to survive in the meantime. Although I did not yet have a name for this suffering I was learning to endure, it was very definitely my first prolonged experience of the Trouble with Paradise, and caused me to become quite reflective.

Perhaps, being part of a church going family, I would have developed a sense of spirituality in any event, but perhaps such reflectiveness also helped me become more sensitive to this side of myself. Apart from being hyperactive, I was also hyperlexic, meaning I had been reading voraciously since the age of five, with the help of a photographic memory. I had overheard my parents discussing me one evening, and using the word "psychologist". Carelessly, they had left a child psychology book on the bookshelves, and I started to read .... At the age of eight I started asking awkward questions about God, the Universe, evolution and other things I had been thinking about.

Luckily my father was also an omnologist (a word coined by author Howard Bloom to describe himself)—someone with an encyclopaedic memory. When I asked him if the universe ever ended, he explained that infinity was like one giant inside the tummy of another giant inside the tummy of an ever bigger giant, and so on ad infinitum. While I mulled this over frequently, his answer did not seem to be fully satisfactory. From my frequent sessions in church and Sunday school over the years, I had begun talking secretly to God through my own direct channel, and I did feel some kind of "presence" around or near me from time to time.

I have made it my business since then to fully investigate in every way possible the nature of religious and spiritual experiences, both in terms of my own experiences as well as those of others. This has involved both following my own intuitions, as well as sharing such things with others, and learning new methods and practices at every turn from those whose credentials I respected most highly. Most importantly, I learned to distinguish between teachers who walk their talk, and those whose actions and lives do not match their words.

Thanks to the liberal, progressive Canadian system, by the time we left Canada to return to South Africa at the end of 1968 I had developed a workable balance between the early stages of the Affiliative, Human Bond "Green" developmental niche, and Achievist "Orange". Despite my earlier troubles with the bullies, I had become "one of the gang" with my youthful rock/pop band, and winning the athletics championships a few years in a row also boosted my confidence. Life was looking very good indeed. I left at the top of my class, with a wonderful group of friends.

I was in for a rude shock to my sensibilities, but it would take a while to show up.

Back in South Africa, the economy had boomed, and since the arrest of the main ANC leaders political tensions had eased slightly. What had not changed was the mindset and culture of the country, which was stuck in the 1950's. In particular, the high school and education system was still modelled on the traditional British private school. If you've read Harry Potter books, you will remember there were four houses into which one was sorted into upon arrival. Check. There were prefects, whose job was to maintain the discipline and ethos of the school. Check. School Uniforms. Check. A great deal of emphasis was placed on school sports, winning and "school spirit". Check. There were regular haircut and personal hygiene inspections, to ensure that long-haired, rock 'n roll values and short skirts remained outside the school gates. Check. Corporal punishment. Check.

After the open book exam, open classroom, liberal Canadian educational experience where we were all a super green team, this slightly fascist, hierarchical model came as both a shock and an unpleasant surprise to me upon my beginning my first term at Bryanston High School. Luckily there were some countervailing forces at work, including our liberal parents, and some very progressive teachers who made their classes a pleasure.

The highlight of my life during those first high school years was the "Raspberry Traffic Jam", our rock 'n roll band. We fully embraced all the countercultural musical trends, and wore bellbottom trousers, Mao collars and flared paisley shirts. Our parents were not amused, but the sixties and early seventies was beginning to wash over the world, and Bryanston, Transvaal, South Africa was to be no exception.

On the one hand the countercultural revolution of the sixties and seventies was an Affiliative "Green" Reaction to the excesses of the Achievist "Orange" system: pollution, stultifying boredom at work, conformity, materialism, the Military Industrial Complex and war, alienation and emptiness, and a worship of science and technology above human values. With a strong dash of Romanticism thrown in, us teenagers were being encouraged to "Tune In, Turn On and Drop Out", while at the same time going to school in a uniform and having to obey a system which seemed headed in the wrong direction. It was all rather heady, yet very confusing at the same time.

Where the counterculture movement in America had the Vietnam war as a rallying cry, in South Africa we had apartheid. At our school of a thousand people, there were really only two people who seemed to stick their head above a parapet against apartheid: my friend Philip and I. Phil's dad was the quietly spoken head of the Lutheran church in Southern Africa, a friend of Bishop Desmond Tutu, and a steady campaigner for rights for all. We would get ourselves into trouble as soon as politics was discussed, and be forced to shut up by the adults present. On the surface, things seemed to be going well in the Beloved Country, so why stir the pot?

The youth club at the Lutheran church I began to go to with Phil was also full of progressive liberals who were against apartheid, and I found my teenage "tribe" amongst this small minority group in a country run by fascists for and on behalf of the capitalist mining houses, and other major employers of labour. Apartheid was above all an economic system, designed to provide cheap labour for the mines, farms and industry, an uneasy alliance between right wing conservative politicians who believed that the Afrikaners were God's "chosen people" in Africa, and socially liberal but economically conservative business people who could see no alternative if they wanted to make good profits.

My father's ambitions for me were clearcut: to become a successful businessman and sportsman, like he was. While I excelled at gymnastics and athletics (and skiing in Canada), I was never very good at ball games. My superbly gifted ball game-playing father worked hard coaching me to become a great rugby player, golfer, tennis player, squash player and more. By my mid teenage years I had been coached in and played around 30 different sports, and was not particularly brilliant at any of them.

While I learned much about team dynamics and strategy through these efforts, and expended a great deal of youthful testosterone smacking or passing a variety of different sized balls in many different directions, at the end of the day I derived almost no lasting personal satisfaction from any of this effort, apart from being superbly fit and muscled. This British Empire derived emphasis on sports was designed to build strong bodies and healthy minds to make young men ready for Empire building. As I was coming of age, however, Empire building was going out of fashion, and Britain was giving away all of her colonial possessions. This egocentric "Red" adolescent energy was going to have to be channelled in a different direction.

Thanks to the Canadian education system, I had learned to speak my mind without fear, often in front of large audiences for speaking competitions. At the age of 15 I was invited to participate in the election of Junior Town Councillors for the Sandton[16] Mayor's office, got elected to the Council, then in turn was elected Junior Mayor of Sandton. This opened up a whole new world to me: the world of politics, and I loved it.

My duties included flying in with the Junior Mayoress to open the Bryanston Shopping Centre by helicopter, speaking at public functions with the Mayor, and raising money for charities. The latter involved setting up folk, rock and pop concerts all over Sandton, a "Battle of the Bands", and a variety of other activities devoted to keeping Sandton "clean, green and serene", my campaign slogan. The poor in the black township of Alexandra were also beneficiaries of the funds we raised, and there was a wonderful win/win/win atmosphere in everything we did. We were pre-eminent, having fun, making money and making a difference—a theme that has stuck with me throughout my life. The ten thousand indigenous trees we planted throughout Sandton in 1972 are today maturing into a beautiful canopy of gracious celtis Africana (white stinkwood), which along with all the other trees planted over the decades, have made Sandton the world's largest manmade forest.

I learned a powerful lesson at the age of 16 that has guided much of my activities since then: that no matter what your age or circumstances, you can make a positive difference in the world if you put your mind to it. I also began to see that one could create safe and productive spaces and places between the polar opposite forces that were threatening to tear South Africa apart, and which would be resolved in a largely peaceful revolution culminating in Mandela's election as President in 1994.

A year or two after psychologist Professor Clare Graves gave his speech to the American Psychological Association on the "great leap" in the minds and values he saw emerging in his data on his thousands of college students in New York, something wonderful was dawning in my own mind as I left school and started my first year at the University of the Witwatersrand in Johannesburg in 1974.

---

[16] Sandton was a new town recently formed out of **Sand**own and Bryans**ton** in the early 1970's. Today it has become the business capital of Southern Africa.

The central theme of Professor Graves' 1970 speech to the American Psychological Association is encapsulated in the following quote[17]:

"The present moment finds our society attempting to negotiate the most difficult, but at the same time the most exciting transition the human race has faced to date. It is not merely a transition to a new level of existence but the start of a new 'movement' in the symphony of human history".

What exactly was Graves picking up on in his extensive two decades worth of research data the enabled him to make so bold a statement? In an article published in the "Futurist" magazine in 1974, Graves laid out in great detail why he believed "Human Nature Prepares for a Momentous Leap". In the following paragraphs he outlines what his data was telling him:

"For many people the prospect of the future is dimmed by what they see as a moral breakdown of our society at both the public and private level. My research, over more than 20 years as a psychologist interested in human values, indicates that something is indeed happening to human values, but it is not so much a collapse in the fiber of man as a sign of human health and intelligence. My research indicates that man is learning that values and ways of living which were good for him at one period in his development are no longer good because of the changed condition of his existence. He is recognizing that the old values are no longer appropriate, but he has not yet understood the new.

---

[17] Don Beck and Chris Cowan describe the seminal work of the late Professor Clare W. Graves, Union College, New York in "Spiral Dynamics". Graves described what he called "Levels of Psychological Existence" as an emerging pattern and priority of worldviews, value systems, and complex adaptive intelligences that arise in response to Life Conditions. Thus, human nature is not finite. We are not frozen into types or traits. Cultures are not static entities, forever trapped in Flatland. As Graves explained it: "Briefly, what I am proposing is that the psychology of the mature human being is an unfolding, emergent, oscillating, spiralling process marked by progressive subordination of older, lower—order behaviour systems to newer, higher-order systems as man's existential problems change."

The error which most people make when they think about human values is that they assume the nature of man is fixed and there is a single set of human values by which he should live. Such an assumption does not fit with my research. My data indicate that man's nature is an open, constantly evolving system, a system which proceeds by quantum jumps from one steady state system to the next through a hierarchy of ordered systems."

I was lucky to find not just hundreds, but also thousands of fellow students who shared some of my values, and to some extent my aspirations, as a first year student at Wits University. I was studying law, languages and economics in a Bachelor of Commerce degree, but found I had much in common with the more politically inclined students who were in the main studying the arts, politics and occasionally the sciences. The engineers were generally not interested in politics at all, and if they were, preferred conservative to liberal approaches.

What bound us together in the growing anti-apartheid protests on campus, was both a vision of a much better, fairer future for all South Africans, as well as a disgust with the brutality being meted out by the SA Police and the increasingly active Bureau of State Security and their plain clothes thugs. Yet, we experienced little or no fear in taking on such a rabid police state system, otherwise we would all have "ceased and desisted" our activities immediately.

What Professor Graves had found in his research amongst New York students at exactly the same time mirrored our own experience in South Africa:

"As man moves from the sixth or personalistic level (the Affiliative "Green" value system"), the level of being with self and other men, to the seventh level, the cognitive level of existence (what we now call Authentic Integral or "Yellow"), a chasm of unbelievable depth of meaning is crossed. The gap between the sixth level and the seventh is the gap between getting and giving, taking and contributing, destroying and constructing. It is the gap between deficiency or deficit motivation and growth or abundance motivation. It is the gap between similarity to animals and dissimilarity to animals, because only man is possessed of a future orientation."

It was from this orientation toward the future, and the motivation of growth and abundance, that we shared an excitement about our protests to free Mandela and to build a thriving, prosperous South Africa for all of its inhabitants. While the Integral mindset was emerging in fragments distributed rather unevenly throughout our generation, it was these thousands of "mind grenades" which gave us the courage and hope to do what we did. There was no hatred, and little fear. The future was bright, the future was integral—we just did not call it that at the time.

Professor Graves again puts my own experience better than I can:

"Once we are able to grasp the meaning of passing from the level of 'being one with others' to the cognitive level (Authentic Integral level) of knowing and having to do so that 'all can be and can continue to be,' it is possible to see the enormous differences between man and other animals. Here we step over the line that separates those needs that man has in common with other animals and those needs which are distinctly human.

Man, at the threshold of the seventh level, where so many political and cultural dissenters stand today, is at the threshold of being human. He is truly becoming a human being. He is no longer just another of nature's species. And we, in our times, in our ethical and general behavior, are just approaching this threshold, the line between animalism and humanism."

This was part of the great shift that produced successive waves of change over the next two decades that brought Nelson Mandela to power as the first black President of a democratically elected government in South Africa, a feat which all authorities and experts had pronounced impossible decades earlier.

During this intensely political period in my life from 1972 to 1982, where I went from being Junior Mayor of Sandton to an Executive Member of the National Union of South African Students (NUSAS) and an Advocate of the Supreme Court right at the heart of the anti-apartheid movement, I also underwent a variety of religious and spiritual transformations.

My good Lutheran friend Phil's family went back to America in my last year of school, leaving me without the connections I needed to maintain my own emerging values which were distinctly different from the rest of my more conservative peers. Being Junior Mayor was also isolating in a personal sense, as people related to me primarily in that role and my own personal needs and values were kept in a secret box I dared not share with others.

For a short while I attended an evangelical church, where I at least felt emotionally supported, but intellectually the rather backward anti-science mentality of most of these good people left me cold. I had undergone a religious conversion experience for the first time as well, partially motivated by the end of my first rather innocent teenage love affair with a Swedish girl, whose family also left the country to return to Europe. After a few months I decided to leave "The Assemblies of God", and immersed myself in the works of William James and others who themselves had undergone similar experiences, in an effort to better understand how someone so educated could get caught up in such a fundamentalist, almost cult-like sect. I began to understand the immense power of different forms of "brainwashing", from religion to the kinds of things the Viet Cong and Maoists were up to.

When I arrived at University, my existentialist friends threw me a lifeline: humanism and atheism. I devoured the works of Sartre and Camus and other existentialists, and underwent an intellectual conversion into a post-modernist for a while. By day I studied business, law, mathematics, statistics—by night and on weekends I was engaged in student movements of varying kinds, political, cultural and sporting. I tried acting in the theatre (never going to make it to Hollywood), guerrilla theatre, and a variety of other resistance style activities popular at the time. I read Marx, debated Hegelianism and theories of social change, and generally tried to understand it all. The human potential movement, Transcendental Meditation, yoga—you name it, I tried it[18].

Luckily, unlike in parts of Europe and North America, my life conditions did not allow me to be captured by either the Affiliative "Green" nor the Achievist "Orange" systems. I was firmly stuck in the fascinating high tension field between these two ways of being and doing, and watching a new system emerge as a synthesis of the thesis

---

[18] Apart from drugs—after spending three months of weekends in the medical school library researching LSD and its effects, I was awarded the only 100% mark I ever achieved in my life for my detailed report of a few hundred pages that highlighted the very real dangers of LSD genetically and psychologically.

of Industrial Capitalism and the anti-thesis of "There's gonna be a revolution", as the Beatles put it in their song "Revolution".

Graves captured beautifully the excesses of the "Green" way of being, where it could get away with it, as follows:

"The idea of a future suffered a similar fate. American Achievist man was always insistent that he had a great future, a 'manifest destiny' somehow enhanced by never having lost a war. Therefore, Affiliative "Green" man, in his rebellion, was forced to throw the future into the same garbage heap as technology, erecting in its place 'the here and now.'

Picture, if you will, "Green" man seated in a yoga position, contemplating his inner self. He has completed the last theme of the subsistence movement of existence. There are no new deficiency motivations to rouse him from his meditations. In fact, he might well go on to contemplating his navel to the day of his death, if he only had some suitable arrangement to care for his daily needs. And it is quite possible for a few "Green" individuals to live this way. But what happens when the majority of a population begins to arrive at the "Green" level of existence? Who is left to care for their daily needs? Who is left to look after the elaborate technology which assures their survival? If we return to "Green" man seated in his yoga position, we see that what finally disturbs him is the roof falling in on his head.

This roof can be called the T problems, the ecological crisis, the energy crisis, the population crisis, limits to growth, or any other such thing which is enough of a disturbance to awaken "Green" man. Naturally enough, his first reaction will be that evil technology is taking over and that all the good feeling and greenery which made the Earth great is in the process of being wrecked forever. (We remember that attitude from the days when his father, Achievist "Orange" man, had much the same erroneous notion.) "Green" man is correct in the sense that his entire way of life, his level of existence, is indeed breaking down: It must break down in order to free energy for the jump into the Authentic Integral state, the first level of being. This is where the leading edge of man is today."

At the end of the 1970's my life continued to be poised uneasily between the realities of making a living as a newly qualified lawyer working for the top law firm in South Africa, while also seeking to my continue my social activism. Without knowing it, in all my unease and concern about the future of my country and the world, I was navigating what Clare Graves so eloquently called: "The leading edge of where man is today", and it was a dynamic but rather uncomfortable, chaotic place, full of surprises and sudden changes of fortune.

The next rude shock awaited me: two years of national service in the South African Defence Force ("SADF"). As a student I had been able to delay the draft, but in 1979 I was called up to serve in the Army during 1980 and 1981. At the time, South Africa was effectively engaged in five wars on five very different fronts, with clandestine support from the CIA and "anti-communist" right wing agencies around the world:

- **Mozambique**—the communist resistance movement Frelimo had toppled the corrupt Mozambican government left behind by the Portuguese. Led by the late Samora Michel (whose wife Graca was subsequently to marry Nelson Mandela), Frelimo posed a real threat to the north-eastern parts of S Africa. One million people died in the fighting and five million were displaced.
- **Rhodesia/Zimbabwe**—Robert Mugabe's terrorists were gaining ground against the white controlled government of Rhodesia, which would be extracted from a bloody civil war by the Lancaster House accords. Zimbabwe enjoyed a few prosperous years after Independence later in the 1980's, then headed down the road to disaster led by Comrade Mugabe to become a failed state in the 1990's. Miraculously, "only" tens of thousands died in this viciously fought war.
- **South West Africa/Namibia**—Sam Nujoma's SWAPO terrorists had gained substantial ground against the SADF, and eventually SA handed Namibia over to SWAPO in "free elections".
- **Angola**—a civil war raged in Angola for over two decades between the MPLA and UNITA, while the war served as a surrogate battleground for the Cold War, engaging the Soviet Union, Cuba, South Africa and the United States. By the time the MPLA won, 500 000 people were dead and one million displaced.
- **South Africa**—the ANC and PAC terrorist campaigns were hotting up inside S Africa, as well as conflict between the Zulu Inkatha Freedom Party and the Xhosa dominated ANC. It is a small miracle

that "only" tens of thousands died in this internal struggle, relative to the millions in neighbouring countries. Thankfully the country did not end up being called "Azania".

The year I entered the Army, all five conflicts were at their peak, and I was trained in counter-insurgency and anti-terrorist tactics, as well as becoming a legal officer. If things looked difficult for the future from "civvy street", going into the Army and visiting the Border areas really opened my eyes to the unbelievable slaughter taking place across the 3 700 km between Maputo in Mozambique and Luanda in Angola.

Despite doing what I could to support those being persecuted by the S African government while in the Army for passive resistance or conscientious objection, the dogs of war were being unleashed in an awful way everywhere I went, and I saw how rapidly civilized, educated men could regress into war hungry savages, into the heart of darkness that lies in Industrial scale egocentric militaristic "Red" aggression. It was much worse than Syria is today in 2014—there were no NGO's or international humanitarian missions to help, just death on a vast, industrial scale. Neither the Americans, Russians nor South Africans were going to give an inch.

During this time I had the opportunity to travel up and down the spiral of development at high speed, faced with situations that only a big, dangerous war can provide. Catch-22, the legendary novel and film, captured some of that insanity:

"There was only one catch and that was Catch-22, which specified that a concern for one's safety in the face of dangers that were real and immediate was the process of a rational mind."

"Just because you're paranoid doesn't mean they aren't after you"
"They're trying to kill me," Yossarian told him calmly.
No one's trying to kill you," Clevinger cried.
Then why are they shooting at me?" Yossarian asked.
They're shooting at everyone," Clevinger answered. "They're trying to kill everyone."
And what difference does that make?"

I began to plan my exit from South Africa during those two very hard, sobering years. My first step was to find a career which was international, and I very luckily found a Diploma in International Financial Management I was able to do part-time in the second year of my military service. The program was run by a dynamic young Vice-President at Citibank, who offered me a superb and challenging job. I left military service in December 1981, and changed my uniform for a suit to start work the next Monday. I had successfully completed military service without having my head blown off.

I was one of the first three people in a unit that became the first investment banking operation for Citibank in Africa, and after being sent abroad for training for several months in London, decided to make that my next port of call. I also had the opportunity to contribute to the Sullivan Affirmative Action program Citi ran in South Africa, tutoring coloured and black staff in economics and writing reports on how our lending did not assist the apartheid government in any of its nefarious activities.

Before heading to London, I took one more job, ending up as the head of Corporate Electronic Banking at the Trust Bank of Africa. I had discovered the power of personal computers and financial modelling at Citibank, and decided that, given I did not want to be an investment banker (those people in red braces you see screwing you and our financial system in the movie Wall Street were already prowling the corridors at Citibank), what better than focus on information technology, the future? After all, Bill Gates had said that he was going to put a PC on every desk, and with the help of Steve Jobs and others, who was I to stand in his way?

My last goodbye to Africa was to fly the two-seater motorised glider I owned a share in, to Victoria Falls and Lake Kariba for a final African holiday. The 1 500 km, ten hour flight took my girlfriend and I over the majestic Okavango swamps in Botswana, which stretched a far as the eye could see. Avoiding the rocket launchers at Katima Mulilo on the Angolan border, we turned right when we saw the Zambezi River and followed its mile wide contours all the way to Victoria Falls, or, as the locals call it, Mosi-oa-Tunya (Tokaleya Tonga: the "Smoke that Thunders"). Almost two kilometres wide and 110 m high, the Victoria Falls are an awe-inspiring sight. The cloud rising above the falls to nearly 800m can be seen from 40 km away, and on the day we flew towards it, it joined with the puffy cumulus clouds above to form a white tower in the sky.

A week later, we landed on a small bush strip at the Bumi Game Lodge, circling the runway to warn the elephants absorbing the warmth of the tarmac at sunset. Reconnecting with the incredible natural beauty and warmth of the local Tonga people in this part of Africa only reinforced the extent of the post-colonial tragedy unfolding on this giant continent. Paradise had been turned into Hell for so many, on so many occasions, one wondered if the true potential of Africa would ever emerge into full daylight.

And so my new life began with me boarding a 747 to London from Johannesburg at the end of 1985. I became an immigrant to the United Kingdom, and officially emigrated from South Africa, followed closely by half a container load of my most precious possessions on a ship out of Cape Town. I did not claim political exile, although I had been closely watched by the Bureau of State Security since I became active in student politics at Wits. Rather, I had decided to create a new future based on three passions: information, transformation and globalisation.

I left just before the civil war in South Africa peaked, (though the bloodshed in the townships continued for another decade until Mandela and the Truth and Reconciliation Commission ushered in a more peaceful era for the country and the whole of Southern Africa). At the time I left the situation seemed hopeless, and everyone who could leave, did. I wanted to see more of the world, and not feel captured by a situation I now realised I had very little chance of influencing positively. I wanted to settle down eventually, have a family and build a career that would last, somewhere stable and with a bright future for my descendants.

At the time I felt quite alone, though I was one of several million people around the world who had chosen to leave their country of birth to create a better future elsewhere, a trickle that began to turn into a flood in the 21st century. Recent statistics show that there are now over 200 million people living in a country other than that of their birth[19], or about 3% of the global population. I took it upon myself as a personal challenge to decipher this mysterious quality of "Britishness", something I had observed from a great distance for most of my life from two very different continents.

The early to mid 1980's saw a major cultural turning point in the developed western countries where Affiliative "Green" movements began to be accepted as part of the mainstream, while Achievist "Orange" governments were elected in Europe and the Americas as part of the backlash against policies which nearly bankrupted the UK and several major American cities. The Great Society dream of liberal and social democrat politicians of the previous decades was abandoned as simply unaffordable and impractical, and updated with a more adaptive (yet deeply flawed) approach to the changing times.

Politicians of the time from Ronald Reagan and Margaret Thatcher set a new tone of "responsibility" and "enterprise", placing the market at the centre of political debate, while a wave of privatisation and institutional reforms swept around the world for the next few decades. Along with the rise of the Asian "Tiger" economies, together with reforms in China and India, new sources of low cost production opened up that enabled three decades of outsourcing of jobs and facilities to low cost producer countries, dramatically diminishing the bargaining power of relatively expensive labour in the developed world.

Many of these reforms were overdue, but many of them were also overdone. Big investors used the opportunity to "make a killing" with

---

[19] As of 2006, the International Organization for Migration has estimated the number of foreign migrants worldwide to be more than 200 million. Europe hosted the largest number of immigrants, with 70 million people in 2005. North America, with over 45 million immigrants, is second, followed by Asia, which hosts nearly 25 million. Most of today's migrant workers come from Asia. In 2005, the United Nations reported that there were nearly 191 million international migrants worldwide, about 3 percent of the world population. This represented a rise of 26 million since 1990. 60 percent of these immigrants were now in developed countries, an increase on 1990. Those in less developed countries stagnated, mainly because of a fall in refugees. Contrast that to the average rate of globalization (the proportion of cross-border trade in all trade), which exceeds 20 percent. The numbers of people living outside their country of birth is expected to rise in the future. Source: Wikipedia.

the help of their red-brace wearing friends in investment banking and politicians keen to attract foreign investment. The seeds of increased disparities in income and wealth were sown, leading to three decades of regression to the mean for the incomes of those in the middle classes, while poverty rates began to climb again in many developed countries.

My own agenda had been inspired by two very contrasting thinkers and writers: on the one hand there was American management guru Tom Peters whose calls for innovation, entrepreneurship and liberation made a great deal of sense for those with the ability to grasp the opportunities opening up in this rapidly globalising world. On the other hand, I was also enamoured of a witty yet acerbic Canadian professor of Management, Henry Mintzberg, who was highly critical of the "MBA machine" at business schools, and concerned about the future of a world in which the manipulation of numbers and financial figures would dominate over the ability to create things of lasting value through deeper, more thoughtful approaches to enterprise and management in general.

The career I chose to pursue my triple passion of information, transformation and globalisation was perfect: I became an information technology strategist at PA Computers and Telecommunications in London in early 1986. After two years of fascinating assignments in London, Bermuda, San Francisco and New York, one could already see the transformative power of information technology at work in financial institutions, though their cultural transformation was still clearly lagging their technological advances.

I had also enrolled part-time in the doctoral programme at the London Business School, to research the impact that information technology and globalisation was having on organisations in general. The potential for this technology to help mankind develop a "global brain" was clear, and potentially transformative, no matter what fuelled the early stages of its development. My own enthusiasm for and researches into artificial intelligence ("AI") since the mid 1980's, including playing with expert systems programming and neural networks, convinced me that just as machinery had begin to remove the drudgery of heavy manual labour since the dawn of the twentieth century, so too would AI and related technologies help us to remove boring and repetitive tasks from the white collar workplace.

When I was offered the role of head of Marketing and Planning at the London HQ of NYNEX's BIS Group[20] in late 1987, I jumped at the opportunity to help integrate a leading American telecommunications provider with a software and computing services company with operations around the world. I was able to see at first hand the workings of a dynamic global firm involved in directly transforming the operations of major organisations through their advice, software and services. Working at Board level in a well run, relatively enlightened business, also inspired my own entrepreneurial ambitions, and in 1990 I left to start my own IT advisory firm, and to run the world's first conference on Transformation.

Though my love of philosophy and psychology started early in life, philosophers such as Jean Gebser, Ken Wilber, and psychologists such as Clare Graves and Don Beck had not made it into the mainstream by 1990. Since reading the "Greening of America" and much Bertrand Russell as a teenager, along with the existentialists and post-modernists at university, I had been inspired by futurists such as Alvin Toffler and writers such as Marilyn Ferguson who talked about a "paradigm shift". Books such as the "Strategy of the Dolphin" also suggested we were moving as a civilization from the reptilian brain stem and the limbic system, to a species where our neo-cortex became the centre of control for our being and behaviour.

My PhD research began to uncover a quite remarkable shift going on in large organisations, in a range from what appeared to be a Conformist "Blue" baseline through a strong Achievist "Orange" core, through to a few examples of Affiliative "Green" taking root in parts of the most advanced organisations. And of course there was also a fair amount of Egocentric "Red" lurking beneath the surface of some of the less developed organisations. The "paradigm shift" Toffler, Ferguson and the Strategy of the Dolphin authors were talking about was clearly taking place from a Conformist "Blue"/ Achievist "Orange" baseline up through an Affiliative "Green" system to a place where systems thinking, viable systems research and many other approaches typical of integral and second tier thinking and values systems. In fact, this

---

[20] NYNEX subsequently became Verizon and the BIS Group was sold to ACT which was subsequently integrated into the Misys Group, a global software provider.

very strong concern for world futures was typical of the value system Professor Clare Graves had been describing in his speech to the 1970 American Psychological Association, and which he was picking up for the first time in the late 1960's, just as Marshall McLuhan began talking about the "Global Village".

During the course of my research I compiled extensive statistics on the performance of 24 global organisations over a decade, and linked that into a database I created on the business environments, strategies, business models, leadership styles, organisational cultures, management processes and systems and performance measures of these multinational creatures. I followed this up with extensive interviews and questionnaire completion exercises on these topics with 72 of the Board directors of these organisations, over two years, including CIO's, CEO's, CMO's and CFO's. The organisations were deliberately selected across a wide range of industries, and the only thing they had in common was that their head offices were in the UK and they were amongst the "UK Top 500" organisations.

More by accident than design, the framework I created for the analysis of all of this data and first hand observations, showed the exterior, objective qualities of the organizations on the right-hand side of the "Wheel Diagram", and the interior, more subjective qualities on the left-hand side, which turned out to mirror the lower left hand cultural and lower right hand socioeconomic systems quadrants in Wilber's integral model.

The whole point of Wilber's "All-Quadrant, All-Levels" model, commonly known as "the AQAL model", is that no system can function successfully if one or more quadrant is neglected while others are favoured. This was exactly what my research was demonstrating—that those firms that balanced their development across the four quadrants were much better performers, and much more sustainable over the longer term, although it was to be a few more years yet before I was to discover the AQAL model for myself.

This lack of balanced development was what I called "misalignment" between the eight wheel variables. The main conclusions of the research were that:

- there is a strong correlation between long-term strategic performance in a business, and the level of alignment between the eight wheel

variables. In particular, the alignment between the business environment, business ecosystem and the other six variables was critical.
- there are several different ways in which organisations can successfully manage change, depending upon their specific situations and environments
- there is tremendous scope for improvement in the process of alignment between the leadership style, organisational values, management processes, knowledge/information management & performance measures and management tools and systems in an organisation. The alignment process was messy, inefficient and often ineffective, as it was not consciously managed according to a coherent agenda by the management team
- there is a need for more effective ways of managing the process of business evolution, and for tools which can support the change management process in the often unique circumstances organisations find themselves in.

It was clear from all the evidence that although the culture and climate for adaptive, collaborative knowledge work was being championed by a number of the organisations surveyed, they were experiencing great difficulty in changing the mindsets and behaviour of both the management team and people in the organisation. There were several other barriers to achieving their goals of becoming "customer focused, quality driven, process managed learning organisations":

a) the current organisational structure often hindered collaboration or made it impossible
b) short-term performance pressures made the creation of a learning environment difficult, and resulted in risk averse behaviour which was not conducive to innovation and entrepreneurship
c) there was a lack of vision and leadership at the top, or too much vision at the top without other visions being shared or communicated
d) information and management systems were barriers to sharing knowledge and information
e) functional stovepipes focused too narrowly on specialised roles and tasks rather than broader competencies and capabilities

f) management processes and workflow was fragmented with islands of automation
g) budgeting and planning processes resulted in territoriality.

Most of the 72 directors who participated in the research felt that there was a great need for a more coherent management framework, which would enable them to see the "wood for the trees" in managing the complexity of adaptation and transition in their organisations. Since the research was completed, several major multinational corporations have used the wheel of management as an integral part of their change management process to great effect[21].

What most excited me about these research findings, was the clear potential in all organisations to evolve into better corporate citizens and better places to work, in addition to them being strategically and financially successful. Yet without the latter, the former was not possible, a simple truth much neglected by the Affiliative "Green" value system and politicians whose skill appears to be mainly in distributing rather than creating value. A sad case in point as I write this is the spectacularly unsuccessful socialist French President Francois Hollande, whose policies continue to prolong the misery of the French people and their ailing economy. One commentator recently quipped that it would be nice if President Hollande had appointed at least one of his eight senior ministers as someone who had created a job in his life[22]. As they say in France: "Plus ca change, plus c'est la meme chose", which roughly translates as: "The more it changes, the more it stays the same".

I learned a few important things from the conference on Transformation I ran in 1990, with eight well known speakers featuring the famous, and very funny, Dr Henry Mintzberg. Firstly, no one knew how to define Transformation, especially in an organisational sense. In my presentations, I was using the analogy of the butterfly emerging

---

[21] For a much more complete treatment of the research that led to the Alignment Wheel, see chapter 4 of my book "Managing Complexity", Economist Books, 2000.

[22] The number of people in France dissatisfied with Hollande and his policies reached 79 per cent in early 2014, an all time record for a French President, according to the poll conducted every month by the Institute for Social Studies for the newspaper "Journal du Dimanche".

from the chrysalis, and I now know that the "Est" movement (which subsequently transformed into the Landmark seminar movement), had defined Transformation as the "genesis of a new realm of possibility". In 1990, however, transformation was beginning to be a label management consultants could attach to business process re-engineering programs, which did not imbue it with the transcendent qualities I believed were inherent in the concept.

What was worse, the generally beneficial quality movement was being hijacked as "process management" and then "business process re-engineering", and its most colourful proponents such as Michael Hammer, were talking about "cutting the fat out of corporations, and frying it". In a short book I co-authored at the time critiquing this "blood on the carpet" approach[23], we argued that designing organisations "as if people mattered" would be a much more productive, sustainable approach, and decried the short-term, quick buck emphasis of the (mainly) American consulting industry. Over a century ago even Henry Ford had figured out that his workers should not work more than 40 hours a week, as his own research showed that their productivity would decline and their error rate go up after that.

Luckily, at much the same time, it was becoming clear that customer satisfaction was most highly correlated with employee satisfaction, (in addition to quality of product or service), so the countervailing trend of employee empowerment was on the rise. With the rise of information systems everywhere, the "information society" was on the move and on everyone's lips.

After the Transformation conference, I renewed my efforts to understand this elusive phenomenon, and realised that I would have to dig deeper into the invisible "cultural" aspects of organizations. This began my professional journey to uncover what the internal structure

---

[23] The book was called: "Tactical Re-engineering for Rapid Results", Strategic Directions publishers, Zurich, 1994 by Robin Wood and Giles Taylor. The publisher insisted we use this title, as we were commissioned to write the book, but we would much preferred to have called it: "Re-engineering as if People Mattered—A Modular Approach to Organizational Design".

of the life, the universe and everything is[24], after a number of years dabbling in this area and writing papers on the topic.

During the 1980's I made a habit of reading as many science books as I could—especially where they framed the leading edge, written by authors such as Fritjof Capra, Marilyn Ferguson, Lyall Watson, Rupert Sheldrake, Paul Davies, Stephen Hawking, Roger Penrose and many others. While it is plausible today to assert that the universe is made of information interacting with matter and energy[25], this was a wild hypothesis in the 1980's. I remember attending a book launch for the 'Internal Structure of the Universe', and crafting my own theory of informatics and evolution at LBS in 1987. When I unveiled the equation:

$$E = mc^2 = \log k.w$$

one of the professors said: 'You are either the next Einstein, or you are mad, and I am not sure which!"

"I can assure you I am not the next Einstein, I find advanced maths very tricky", I replied—and dropped the enquiry to get on with my doctorate. The third term in this equation represents the most basic definition of information available, framed in terms of fundamental physics. W = the number of ways in which any system can be arranged, and k = Boltzmann's co-efficient of thermodynamics. Thus, the more intricately arranged a form is, the greater the amount of energy it contains.

Today this would fit well with string theory, where the hidden 7 dimensions wrapped up in the visible 4 dimensions would explain the incredibly amounts of energy we have discovered in so-called 'empty space'. The confirmation of the existence of the Higgs boson in the large hadron collider at CERN in 2012 and further revelations from dark matter research will, I believe, take us another step toward understanding the internal structure of the universe, from the interior of which what we used to call our souls emerge, and to which our consciousness may arguably return upon our death. Ever more sensitive instrumentation is

---

[24] Tom Stonier's book, "The Internal Structure of The Universe", was published by Springer Verlag in 1987.

[25] http://r2meshwork.ning.com/profiles/blogs/the-universe-is-a-quantum-computer

The Trouble with Paradise

now enabling us to detect the subtler frequencies in living energy fields, and to practice energy/information healing on a scientific basis.

It was a quarter of a century ago that I first encountered the architect, anthropologist and information scientist Max Boisot through his book 'Information and Organisations', as I began my PhD at London Business School in 1987. I was immediately in awe of what his "information space cube" could explain about the world in general, from biology through to markets and societies and the world of organisations in particular.

I first met Max in the flesh on a wintry day in Coventry in the early 1990's where the LSE Complexity group were all attending a Complexity and Organisations symposium. Max's endearing wit and brilliance were in full flow. Thus began a decade of friendship at a distance marked by the episodic way in which we kept on bumping into each other and having absolutely brilliant, side-splitting conversations about life, the universe and everything.

When I read Max's first book, 'Information and Organisations', I finally found a process that described how structure or form could emerge from information. The i-Space theory applies to biology, psychology, organisations and societies. Of course today we also understand that cellular informatic feedback processes are as formative as the DNA/RNA/protein process we have believed to be the only key since Crick and Watson and Darwin, and we have discovered that DNA is emitting photons as a part of the coherent structuring process that enables life to exist at all. Energy medicine and mindbody medicine based on this new understanding of DNA vibration frequencies and cellular vibration frequencies will be the next major revolution in medicine, in conjunction with the breakthroughs in the past decade in epigenetics.

This very much describes the frontier of science and spirituality today—do we live in a dead, random universe, or a living, conscious universe? And what exactly is consciousness anyway? I still dream of the day when we can demonstrate mathematically how physical form emerges from the 'zero point' at the bottom left hand side of the i-Space cube, equivalent to the upper left quadrant in Ken Wilber's AQAL integral model[26]. I suspect I am going to have to wait a while for the next set of breakthroughs in physics beyond where we are today to help us get to the bottom of this.

Meanwhile, I was fortunate to meet an amazingly talented group of people in the early 1990's with whose help I was able to make a living and explore my fascination with information, transformation and globalisation. I started off the 1990's advising senior executives on the transformative role of information systems and the need for the alignment of culture, strategy and business models to ensure that their systems helped their enterprises succeed in sustainable ways. I met some of my future business partners through their facilitation of my Transformation conference, which resulted in us co-founding an organization based on the revolution in cognitive science and its applications for management.

The central idea of "Idon", was that of: "**Idea** + **Icon**= Idon". While icons have transformed the world of computing by enabling us to represent functions and information with recognisable symbols (think of the "Trash Bin" icon on a computer where you can drag and drop files), the purpose of Idons is to enable us to represent ideas and thoughts within symbols, and to play with those ideas and thoughts. A whiteboard full of idons becomes a representation of our mental maps and models enabling us to represent our thinking, and also to dig deeper into our assumptions and the things we take for granted. Idons, therefore, turn out to be an incredibly powerful tool to facilitate creative thinking and organizational learning[27].

---

[26] I have tested out this correspondence between the AQAL quadrants and Max's i-Space model with other fellows of the Integral Institute, and found there to be a clear formation cycle whereby ideas emerge in the upper left quadrant, flow across into conscious, visible form on the upper right hand quadrant (codification), then are shared in the bottom right hand side quadrant (diffusion). Once an idea such as '$e=mc^2$' enters the social realm, it then gradually becomes enculturated (abstraction) as part of our culture in the lower left hand quadrant. Our culture in turn then reproduces this knowledge and enculturates it into individuals in the upper left hand quadrant with each successive generation.

[27] See **Hexagons for systems thinking—Anthony M Hodgson**—First published in the European Journal of Systems Dynamics Volume 59, No 1, 1992.

At about this time, Arie de Geus, the head of planning at Shell (who was responsible for the famous Shell scenario process), was looking for a way he could accelerate Shell's learning through scenarios in the company's operating divisions around the world, which resulted in a number of projects for our fledgling firm. As Arie used to say, the big question for Shell (and any other oil company), is what happens when the oil runs out[28]. Arie was also one of the catalysts that helped Peter Senge write his now famous book on Organizational Learning, which played a central role in my work from the moment I read it in 1990.

I first met Arie on a project Idon was doing with Shell in 1992, working with the World Bank to develop sustainable energy lending policies, and was immediately impressed by his gravitas and wisdom, and his deep concern for people and the planet. He was a consummate professional, and we called him "the magician" due to his ability to bring unlikely, very brilliant people (they were called "remarkable people" in our work) together to solve what we now call "wicked problems". Climate change is a wicked problem, for example, because it requires many stakeholders with very different, often opposite interests to collaborate for the common good, without necessarily having any incentives to do so.

I had met a few "wise men" in my life, and looking back on it now, realise that Arie[29] was one of the first examples of transcendent consciousness (or "Turquoise") I had met in business, and that this is what enabled him to be such a magician, pulling off the seemingly impossible in a company with as many constraints placed on it as Shell did and still has.

---

[28] In the 1990's, Shell began to invest in renewables and focus in on sustainability as a key issue, led by Mark Moody-Stuart who became a Managing Director of The Shell Transport and Trading Company, p.l.c. in 1991 and was Chairman of Royal Dutch/Shell from 1998-2001. Sadly much of this good work was undone by executives who took over from Moody Stuart. He was succeeded by Sir Philip Watts, who was forced to quit due to his role in the over statement of Shell's reserves.

[29] Arie de Geus's book: The Living Company, Harvard Business School Press, 2002—"is the genuine expression of a remarkable man".—*Francisco Varella, Director of Research, National Centre for Scientific Research, Paris.*

By 1994 Idon had grown dramatically, having completed dozens of major projects with global organizations and had also hired a few dozen people. As seems to be the case in my glorious career, a few years of success were always followed by a few lean years, and 1994 proved to be no exception. With work drying up from our traditional clients, many of our new consultants found themselves unemployed, so we began a new firm with a new focus called Genesys, which had two meanings: "**Gen**erative plus **Sys**tematic" and "**Gen**etic plus **Sys**tems". Our focus was specifically on Transformation, and we styled ourselves as "Partners in Collaborative Transformation". While Genesys started up, we continued our roles within Idon, though on a much smaller scale than before.

Genesys was an interesting and reasonably successful experiment in a new form of collaborative governance, in which the partners were each rewarded with new shares for generating new business every year, which as anyone can tell you, is the lifeblood of any professional services business. We each got to be pre-eminent, work in areas for which we had a passion and for clients for whom we cared about. And most importantly, we got to continue the innovations that had started with Idon and our own subsequent research into what was going on at the leading edge of Corporate Transformation. It was at this time that I started writing a book which would eventually become "Managing

Complexity—How the New Sciences can Help Businesses Adapt and Prosper"[30].

During the late 1980's "chaos theory" became very popular amongst managers, aided by the success of Tom Peters' book "Thriving on Chaos". Even the renowned, and perhaps greatest management guru of all time, Peter Drucker, made it clear that we were entering an age of turbulence, and chaos theory is all about understanding turbulence in everything from weather systems to fluid flows. Perhaps the most important influence on us at this time in Genesys was the research being

---

[30] Managing Complexity—How the New Sciences can Help Businesses Adapt and Prosper, Robin Wood, Economist Books, 2000. The **Sunday Times Review said: Managing Complexity is "Book of the Week"**—Last year's most fashionable piece of jargon for management consultants was "connectivity". On the evidence of Robin Wood's state-of-the-art survey of complexity management, this year's is set to be "collaborative capitalism"—the idea that, as we move into the "knowledge era" of the 21st century, the old autocratic management styles will steadily become more outdated. In a future where information is king, companies will co-operate with each other as often as they compete, and managerial skill will rest on the ability to fix a balance between the two. If this sounds apocalyptic, it should be pointed out that there is a revolution going on out there. Wood's delvings among the Fortune statistics make frightening reading. The share of global gross domestic product represented by the top 500 is now below 30%. As the gap between economic cause and corporate effect continues to diminish, the average company lifespan across Europe is a paltry 12½ years. It will get shorter still. Offered multiple options and instant technological access, the customer is king. For the seller, staying afloat will require not just better products, but better relationship management. Lose a customer once and you may never pull him back. But this is an optimistic as well as an alarming book. Fortunately, as Wood is keen to point out, our contemporary "macroscope" of connected economic units shows a way forward: "connect and create" organisations will take the place of old-style monoliths. Certainly, many of Wood's case studies of complexity management at the coalface are impressive. E-Tech, for example, sees off its competitors by way of a complicated simultaneous investment in new technology and process re-engineering

carried out at the Santa Fe Institute[31] or "SFI", by leading scientists and philosophers such as Murray Gell-Mann (two Nobel prizes in physics), theoretical biologist Stuart Kauffman (a MacArthur Fellow), economist Brian Arthur and business philosopher Howard Sherman.

Research at SFI focuses on systems commonly described as complex adaptive systems. Recent research has included studies of the processes leading to the emergence of early life, evolutionary computation, metabolic and ecological scaling laws, the fundamental properties of cities, the evolutionary diversification of viral strains, the interactions and conflicts of primate social groups, the history of languages, the structure and dynamics of species interactions including food webs, the dynamics of financial markets, and the emergence of hierarchy and cooperation in the human species, and biological and technological innovation.

What was the relevance of all of that massive intellectual horsepower and those ultra-sophisticated computer models in business, life and the planet in general?

In the 1990's it was becoming clear that the traditional ways of managing and leading organizations was fast becoming a major liability not only for the organizations themselves, but also for our society and our planet. The pursuit of short-term profit, shareholder value and stock price gains through the manipulation of quarterly reports to the stock market had reached a peak.

---

[31] The Santa Fe Institute (SFI) is an independent, nonprofit theoretical research institute located in Santa Fe (New Mexico, United States) and dedicated to the multidisciplinary study of the fundamental principles of complex adaptive systems, including physical, computational, biological, and social systems. The Institute consists of a small number of resident faculty, a large group of "external" faculty, whose primary appointments are at other institutions, and a number of visiting scholars. The Institute is advised by a group of eminent scholars, including several Nobel Prize winning scientists. Although theoretical scientific research is the Institute's primary focus, it also runs several popular summer schools on complex systems, along with other educational and outreach programs aimed at students ranging from middle school up through graduate school. The Institute's annual funding comes from a combination of private donors, grant-making foundations, government science agencies, and companies affiliated with its business network. The 2011 budget was just over $10 million.

The Trouble with Paradise

Meanwhile, in Rio de Janeiro, at the first Rio Summit in 1992, the international community adopted Agenda 21, an unprecedented global plan of action for sustainable development. It was clear that businesses were lagging in their efforts or even willingness to seriously engage with the big challenges that lay ahead.

Complexity makes it hard to forecast the future. Not only are forecasts uncertain, the usual statistical approaches will likely underestimate the uncertainties since key drivers like climate and technological change are largely unpredictable and may change in non-linear fashions. During the 1990's, in an accelerating wave of interest and enthusiasm, complexity-based approaches to organisations, their environments and operations began to spread rapidly throughout business and government. Firms such as AT&T, BT, HP, Sun and others in the computing and telecommunications sectors; manufacturing organisations such as GM, Deere, Unilever and United Distillers; financial services providers such as Citicorp, Barclays, NASDAQ; the defence sector and others in local and central government—all were exploring and using concepts, frameworks and tools derived from complexity.

The field of complexity also offers several ways of tackling the unsolved challenges of management, strategy and organisations. It helps executives and policymakers think about, predict the behaviour of and run business and management processes. Financial traders, computer programmers, soldiers and artists—people from all walks of life are now applying complexity approaches to their own fields of endeavour with great success. Developing a basic understanding of complexity-based approaches, frameworks and tools became to be seen as essential for all executives.

Today the concepts and tools involved in the dynamics of complex systems and their implications for sustainability are developing in parallel, influencing not only the natural sciences but also the social sciences and humanities. Complex systems thinking is used to bridge social and biophysical sciences to understand, for example, climate, history and human action, assessments of regions at risk, syndromes of global change and how to link social and ecological systems for sustainability. It underpins many of the new integrative approaches, such as ecological economics, sustainability science and ThriveAbility.

The complex adaptive systems approach shifts the perspective on governance from trying to control change in resource and ecosystems assumed to be stable, to enhancing the capacity of social-ecological systems to learn to live with and shape change and even find ways to transform into more desirable directions.

I would argue that creative thinking, learning organizations, chaos theory, systems science, sustainability and complexity science all involve parts of the brain that would be conducive to what Professor Graves called "Second-Tier Thinking". That is not to say that Affiliative "Green" types are not capable of doing this kind of thinking from a green values base, but continuing to use these tools would definitely enhance their transition to an Integral "Yellow" stage of development, providing both the life conditions and their own personal practices supported this transition.

It would be naïve in the extreme to suggest that those who are making the great leap from first tier to second tier thinking and values are 100% on the other side of the line between first and second tier, but the real question is where the center of gravity of a person is most of the time. The important thing to remember is that integral thinking can be embedded in either Achievist "Orange" or Affiliative "Green" ways of thinking and being, and that development from one stage to the next can take five years or more to fully shift from one center of psychological gravity to the next.

Idon and Genesys were very fortunate to work with some of the most progressive and enlightened organizations and executives to be found anywhere on the planet during the early and mid 1990's, using some of the most sophisticated methods and tools to diagnose very large complex systems and work with these clients to facilitate major changes that stuck, while also making these organizations more adaptive[32].

In 1995 two new initiatives opened up opportunities to apply our thinking and approaches in two very different worlds: the Achievist

---

[32] Out of the many hundreds of clients of Idon and Genesys, some of the most prominent incude global organizations such as the UK Ministry of Defence, The UK Cabinet Office, The S African Department of Health, The World Bank, 3M, Barclays, BBA Aviation, Citigroup, Eagle Star Insurance, Ernst and Young International, Hewlett Packard, HP/Microsoft alliance, HP/Intel alliance, ICL, Jardine Matheson, Royal & Sun Alliance, Kellogg's, Unilever, Royal Dutch PTT, Shell International, State Street Bank, Vodafone and the ZF Group.

"Orange" world of rising Asian businesses, and the hybrid Affiliative "Green"/Authentic "Yellow" world of Silicon valley.

In the Achievist "Orange" world of rising Asian businesses, I was engaged to help one of the world's oldest corporations prepare for the handover of Hong Kong to China in 1997. I brought to this endeavour every management and leadership tool that had proven effective in my own work since 1975, together with their integration into the diagnostic framework called "The Change Wheel"[33]. Our client also had dozens of under-performing businesses in Asia that needed transformation, so we integrated all of our strategy, organizational development and change tools and methods into a specially adapted package that helped our client's executives to work with us to learn how to do this for themselves.

In real-time, we engaged with value systems right up and down the spiral of development, from the Tribal "Purple" Philippine cultures through the Egocentric "Red" triads and entrepreneurs of Hong Kong, through to the Conformist "Blue" Malaysian businesses, all the way up to the relatively sophisticated Achievist "Orange" Singaporeans.

At the same time, my colleagues and I found ourselves in the hybrid Affiliative "Green"/Authentic "Yellow" world of Silicon valley, where corporations from HP to Intel to Microsoft and Nokia were all pioneering advanced approaches to management and to co-creating business ecosystems. Lew Platt, then CEO of Hewlett Packard, the computer systems manufacturer and IT provider, was without a doubt a second-tier, authentic "Yellow" leader who walked his talk, and it was a real pleasure to be able to work with an organization that attempted to live up to such values. In fact, there was no doubt that HP's values as embodied in the "HP Way", were a major factor in its success and longevity, in an industry where organizations are notoriously short-lived. Jorma Ollila, CEO of Nokia, also shared similar attributes, though both Bill Gates and Andy Grove at Intel still seemed centered in the hyper-competitive Achievist "Orange" way of doing things.

Between 1995 and 1998, we had carried out dozens of projects in Silicon Valley and the USA, as well as a similar number in Asia. From

---

[33] For which I was awarded a doctorate in 1995 by London Business School following a decade of action research in the world's leading corporations, banks and governments.

my base in London, I would be taking phone calls from Hong Kong at seven thirty in the morning London time, and speaking with colleagues in California at seven thirty in the evening. I would sometimes literally fly "around the world" from London to Hong Kong to San Francisco to New York back to London, in order to meet all the commitments in my diary, and this began to take a toll on myself and my young family. Being perpetually jet-lagged was one of the hardest things I have ever had to deal with at a physical level, but it also had consequences for my personal life.

In 1997, Genesys and Ernst and Young entered into a joint venture to promote the strategic thinking and decision making tools we had developed, known as "FutureStep", thus beginning a four year partnership in which I eventually became employed full-time within E&Y. Having helped start up the strategy group in E&Y, my frequent trips to Silicon Valley had also led to us starting an e-business incubator inside E&Y. By the late 1990's, the methods and tools we were using were beginning to spread widely in large organizations, and it seemed as if our generation of business leaders might be on the verge of liberating human potential through organizations everywhere. I even dared to believe that a shift from managerial to collaborative capitalism was underway, as the deafening roar of the dotcom boom engulfed us.

As you can imagine, a life of long distance business travel has both its upsides and downsides. The downsides include being alone, and often lonely, missing friends and family back home, while also being jet-lagged and exhausted from endless journeys via cabs, planes, trains and automobiles. The upside included, for a hyperlexic like me, the opportunity to read and reflect. As I write this, I sit in my office/study surrounded by many thousands of books I bought at hundreds of airports and train stations, as well as from the quaint bookshops around the world I spent many happy hours browsing in. They are like old friends.

On one of my journeys to New York City in the mid 1980's, I had picked up an elegantly bound book called "Inventing the Future" by Marilee Zdenek. Within the pages were a variety of visualization exercises that literally opened up a whole new world to me. To this day, almost three decades later, I still use one of the visualizations frequently it proved to be so effective. I rarely discussed my inner adventures with others, but was fortunate to have had several friends and girlfriends with whom it was possible to share these experiences, and for whom

The Trouble with Paradise

such events did not represent a form of "weirdness" we often associate with those who share everything from their dreams to their revelations without much aforethought.

On every journey I made, I would pick up a couple of books on management, economics, complexity science or strategy, and a couple of books on self-help and spirituality, and devour them en route. In the mid-1990's, I came across the Celestine Prophecy, which though thoroughly "New Age" in its own way, appeared to contain a few nuggets of truth, one of which is that "there are no co-incidences". While such a statement can be carried too far, there appeared to be some truth in it as I examined key events in my own life, and how some of the most impossible things that had happened to me were due to effortless coincidences, while most of my carefully laid plans at which I worked so assiduously, appeared to yield little or no fruit at all. This thoroughly contradicted the "Think Big, Work Hard and Get Rich" philosophy my father and his bookshelves had cultivated in me—how was it that some of the apparently least deserving people, ended up at the "top of the heap"? And some of the most deserving, noble of characters, seemed to be being perpetually kicked in the butt by life?

Deepak Chopra's books have been another big influence on my life, ever since I picked up a copy of "Quantum Healing—Exploring the Frontiers of Mind/Body Medicine" during a seminar on Neuro-Linguistic programming a few decades ago, which several of my colleagues and clients were dabbling in at the time (and which formed part of my first book on "Re-engineering as if People Mattered"). I followed that up with "Ageless Body, Timeless Mind", and then ended up spending a lot of time with "The Seven Laws of Spiritual Success", which came in particularly handy in the many hyper-stressful situations one ends up in in the information, change and transformation businesses. I began to explore what mindfulness and detachment really meant in everyday life, and in organizations. Apart from practicing such disciplines in my own life, I started to meet others who had been influenced by authors such as Deepak, Wayne Dyer and many others, and who seemed to be better off for it.

Given the stress I seemed to be perpetually exposed to, I was also taking any advice I could find that appeared to have a scientific or medical basis regarding my own physical health. A life of business meetings, workshops and travel is a recipe for disaster in terms of weight gain. I had always been hyper-fit and muscular until the age of about 40, when my lifestyle began to catch up with me. Having taken my health and strength for granted most of my life, and being an exercise junkie (my father also did the Royal Canadian Airforce exercises every morning and played sport until the day he died of a heart attack), this was an unusual departure for me. Cue Dr Andrew Weil and others whose books, CD's and advice I assimilated, along with the Blood Type diet and any other advice that seemed to have a chance of success in my hectic schedule.

We all have an inner life, though few are encouraged to pay much attention to it in our modern world. Like most of us, I grew up thinking that my inner life was unusual and somehow not something to be shared with others. It was "private", or even "a secret". Apart from one's closest friends, people were always more interested in gossip, news, sports or hobby talk and interesting facts than something deep and personal.

Part of the problem we all have with sharing anything as confidential as our dreams, hopes, deep concerns, deep emotions, peak experiences or fundamental views on life, is that the risk of being misinterpreted, or even worse, being ridiculed, is high. As the old sign in the bar goes, "No Religion, Politics or Sex Please". These are, and always have been, controversial issues, which raise the risk of conflict, particularly after a few beers or vodka martinis. On the other hand, they also represent some of the most important issues in our lives, which can only be swept under the carpet for so long.

One of the biggest challenges in these "private" areas of our interiors is that we lack a common map of the territory. Imagine trying to navigate from London to Paris if the British and French had completely different mapping systems with no overlap or translation. It would certainly be a great deal more difficult than the stress-free experience we have today with global standards in mapping, navigation, GPS systems, place names and language translators, and the Guide Michelin to find the good restaurants on the way. In fact, just about everything we take for granted in everyday life has been standardized in some way, to make it safer, easier and more affordable—except our interiors.

In this sense, our interiors are the final frontier left untouched by the first Renaissance, and also by successive waves of globalization in the 20th century. It is only in the past few decades that some commonly accepted maps have emerged, and we have touched on them lightly

The Trouble with Paradise

throughout the pages of this book. Every religion, and even atheism, provide us with a set of beliefs; (which in the case of atheism starts with the belief: "There is no God"). Every religion offers rituals. Atheism offers materialistic science—exteriors only. Interiors are for wimps, and psychologists, and if you're really screwed up, psychiatrists. Except for Buddhism, which is atheism with both exteriors and interiors. And rituals.

Which is why it comes as no surprise that the first real "map" of my interior I was to discover (and yours for that matter, though you might like to think that YOUR interior is very special, and quite unlike mine, and of course, you ARE special darling), and all our interiors, comes from a western Buddhist who gave up biochemistry to research consciousness. In an interesting example of synchronicity, I was browsing the shelves in my favorite bookstore in Silicon Valley, Kepler's, just opposite the Stanford campus, where it is de rigeur to believe at least six impossible things before breakfast. I had picked up a great tome of a book called "Up from Eden", by one Ken Wilber. While reflecting on just how big the words he was using were, and how much detail there was, and what practical use I might find in this book (always an important criterion for a pragmatist like me), I heard a nasal German voice talking about epistemology, ontology, the web of life and other things coming through the bookstore speakers.

Curious, as I was not used to hearing nasal German voices lecturing in Californian bookstores, I exited the sky high racks of books to see Fritjof Capra, famous for his "Tao of Physics" and "Web of Life" books (both well thumbed by me already) lecturing to a rapt audience of geeks, students and chic hippies. Still holding my copy of "Up from Eden" (though my arm was beginning to sag under the weight), I listened politely to the very dry, precise and scholarly Dr/Professor Capra until I realized I needed a dose of cappuccino pronto barista.

Truth be told, I already had about a dozen must read books stacked ready for purchase, and Ken Wilber's massive tome did not make it into my shopping basket that evening due to weight restrictions on British Airways, but I did get a personally signed copy of the "Web of Life" for my bookshelves from Dr/Professor Capra. I eventually managed to get a few of Ken Wilber's books delivered to my doorstep back in London, though, through what was then an exciting new startup we all know and love a few decades later as Amazon.

Anyhow, let's get back to mapping our interiors. How is it possible to map human consciousness, when everyone is different? We cannot see inside people's heads. We can map faces, eyes, bodies, diseases and quotients of all kinds, because we have developed elaborate ways of measuring them, with models. I cannot see your IQ, for example, but I can measure it, because we have developed a standard model of intelligence, and experts can give you tests that normalize your IQ against everyone else. So, technically speaking, it should be possible to map human interiors, even if one cannot map their contents exactly. But what models would one use to flesh out what should be in the map?

That was a problem philosophers, psychologists and many others have struggled with for centuries. Wilber took an interesting approach to its solution. He said that one could discern the contours of our interiors from the orienting generalizations made by the world's leading thinkers in philosophy, religion, psychology and other disciplines devoted to matters of the interior. One did not have to believe in any particular religion or subscribe to any particular philosophy, but one could observe the commonalities between them—the orienting generalizations. And if the same phenomena kept on showing up again and again, around the world, over millennia, then perhaps they were important to what it means to be human, and should be on the map.

That sounded like a respectable research strategy to me, which might offer testable hypotheses, much like anthropology and psychology. So I let Amazon take the strain and place many more heavy Ken Wilber books in my mailbox over the next few years. I also had to take my hat off to a man who could spend several years alone in his living room laying out just about every bit of evidence for his map across his living room floor, putting the bits together like some giant jigsaw puzzle. We saw the results of Wilber's efforts in chapter six above, in broad outline, and I find myself using this map of consciousness and the integral model several times a day as I navigate my own life and consciousness, because I find that being aware of the perspectives one is using or "looking through" can make a huge difference to making better decisions and being a happier person. This applies with even greater force to understanding the perspectives others are using.

So it was that a few billion people and I started a new millennium.

The publication of my new book, "Managing Complexity", kicked off the year 2000 with a bang. It won awards, and I became a "Celebrity

Speaker", giving speeches at major events with movers and shakers and famous people on the digital economy, social capital, even meeting with and speaking to Tony Blair's policy advisors a few times. It seemed that most of us who were excited about the e-business boom believed it was the beginning of a golden new era. Valuations of Internet shares continued to skyrocket, and we transformed the E&Y dotcom incubator into a fully fledged e-business unit. Business poured in—the future was digital.

The prevailing rhetoric was that the Internet and mobile telephony would transform everything, and nothing would ever be the same again—business, work, education, science, the military, government and more. And in many ways, the rhetoric proved to be true—as I write this, the Internet and mobile phones have transformed many features of life around the world. In fact around 96% of people on the planet have access to a mobile phone (128% in developed nations and 89% in developing nations)[34]. Over half of those have mobile or fixed broadband subscriptions. Although a PC on every desk and in every home may not be achievable, a smartphone in every hand soon will be.

Most interestingly, we have discovered something that has been in plain sight ever since Narcissus recognized his own face in the still pond: our information technologies and civilization are a mirror of our inner selves. We extend our interiors outward to others through books, poems, drawings, art, blogs, tweets, Facebook posts and every other electronic transaction we undertake. As McLuhan rightly predicted, the thrumming electronic web would weave a new civilization around us, and even in us. We are much more connected and sensitive selves as a result. The transparency of the electronic medium also makes it harder to find places to hide, as Wikileaks so eloquently demonstrated. Knowledge is power, and power is increasingly in the hands of the people.

Of course there are dark sides to this phenomenon, just as there are to every human activity anywhere on the planet. Extremists, pedophiles, spammers, prostitutes, con artists, hackers, bullies, tyrants and the just plain ignorant and opinionated have equal access to the giant thrumming web, though fortunately they are in the minority almost everywhere.

---

[34]   International Telecommunications Union report 2013. http://mobithinking.com/mobile-marketing-tools/latest-mobile-stats/a#subscribers

And perhaps the hardest part: the problem is no longer getting connected, the problem is how to disconnect. People take "Facebook holidays", apologizing to their friends that they are not always "on". We go invisible on Skype, but others still manage to reach us, we feel guilty not answering emails, we end up on a treadmill of our own making without any apparent end in sight. A significant percentage of people take their iPads, tablets and smartphones to bed with them, a truly unhealthy habit, resulting in sleep disruption and unhappy relationships. And all this in less than two decades since the internet first became a public phenomenon, when I was sitting in a room in January 1995 with 100 of HP's top executives in Palo Alto exploring scenarios for the "information superhighway", and Netscape and Amazon had just launched.

For some, an instantaneous apprehension of the totality was becoming possible. From specialists who want to know everything it is possible to know about their field, to omnologists such as myself who are interested in the bigger picture and are curious how everything fits together in this beautiful, wonderfully crazy world of ours, and where it is all going—just about anything of importance or interest can now be found on or through the internet.

Towards the end of 2000, I was headhunted by Scient, a leading global e-business firm based in San Francisco, as a Managing Director in London. E&Y had sold its consulting arm to a large IT services firm, and I could see little future for what I wanted to do there. Thus began a rollercoaster ten months as the dotcom boom turned into the dotcom bust, Scient itself was taken over, and I was given a severance package that gave me some breathing room to figure out what was next for me.

Then, one week after 9/11, after years of heading in different directions, my wife and I agreed on an amicable separation while sitting in our local Starbucks. Separating a family is one of the toughest decisions one can make. Just two weeks later, the third blow struck without warning: my father dropped dead of a heart attack. Yet in the midst of tragedy, life continues. After joining family and friends for my father's funeral in Johannesburg, I returned to London, shaken and deeply saddened, but determined to continue my life's work.

I found a few part-time roles, as a non-executive director and teaching top executives strategy at the London Business School. There were board meetings, presentations and conferences in interesting places in Paris, New York, London, Boston, San Francisco, Monaco, Brussels and Prague. But times were tough, and everything I had invested both financially and energetically into a future nest-egg for myself and my family, disappeared almost overnight.

So began a few years of sorting myself out. The basic questions of life resurfaced: who am I, why am I here, what am I here to do? What is my passion and purpose? What will my legacy be?

I will confess all of this turbulence and such sudden changes in fortune had left me feeling somewhat lost and confused, yet also strangely excited about what my future held. Having an accurate map of my world and myself seemed not only useful, but vitally necessary. I was like an outdated GPS navigation system, using maps from a decade ago in the country of the future, and I was tired of going down dead ends and feeling lost.

I took up an invitation to a weeklong Spiral Dynamics event in Boulder Colorado, which included an opportunity to meet Ken Wilber and a variety of other leading edge integral thinkers. What I needed was a comprehensive update of my personal mapping system. I gained some major insights into who I was, what I wanted and needed to do next, even though none of these topics were on the agenda. As psychologist Steven Pinker says: "Cognitive psychology has shown that the mind best understands facts when they are woven into a conceptual fabric, such as a narrative, mental map, or intuitive theory. Disconnected facts in the mind are like unlinked pages on the Web: They might as well not exist."

My updated map had "You are Here" written all over it, so I was able to begin connecting the dots on my map to find the route to my own small corner of Paradise. I continued to learn and grow during this period, and realised that one of my passions is to mentor and cultivate environments in which others are encouraged to move to the next level in their own development. I also decided to create my own small corner of paradise in the sun, and bought and renovated an old Chateau in Perpignan, the heart of French Catalonia, where I now sit writing these words.

The "Oasis" project was an opportunity for me to roll up my sleeves and engage in a very physical, constructive activity, creating something of lasting value that would outlive me and serve as an inspiration to others. After more than a year of work finding the ideal location, funding the project and beginning the building works with my architect, designer and team, I also realised that it was time for me to find the love of my life, if she existed. I put my profile up on a few online dating sites, and began putting my toes into the water of the online dating game.

While one particular lady accused me of seeking a "personal development princess", I made it clear that one of the priorities and passions in my life is personal and spiritual development. I wanted to spend the rest of my life with someone who valued that in me, and who also valued music, the arts, the outdoors and culture as much as I do. Given the Oasis project was in France, this limited my search to women between 30 to 40 who spoke both English and French, were university educated, and who wanted to live in a small town between the mountains and the sea in the south of France.

After a few initial unsuccessful meetups, blind dates and short-term friendships and romances with a mixture of American, English and French women over a couple of years, I expanded my search to women anywhere in the world rather than just France, the UK and North America. I generally do not believe in miracles in the traditional sense of that word, but at the beginning of 2004 I was about to get lucky, luckier than I've ever been, or deserve to be.

One fresh winter's morning, before the builders arrived, I was browsing the new arrivals on "Friendfinder.com", when out of the corner of my eye I spotted a face that looked both familiar and strangely attractive in an intellectual yet fun kind of a way. I spontaneously penned a short email to "Vintage73", and this began a month of daily correspondence, much of it with our tongue in our cheeks and a mutual sense of humour and the absurd which was both very grounded yet also quite thrilling.

The emails were soon supplemented by daily telephone calls, as we got to know each other in a way that would have been impossible had we met face to face. We had the luxury of getting to know each other's interiors, not just our exteriors. The usual rush of physical attraction often short-circuits this deeper level of communication, yet through our conversations we gradually feel in love with each other. One particular evening, the frustration of being so close to another human being I had never physically met got the better of me, and I offered to meet with her in Moscow where she was working.

The Trouble with Paradise

One early day in spring, I found myself sitting on a sofa in the lobby of the Marriott Tverskaya Hotel in Moscow, waiting for my date. At 2030 precisely, feeling myself to be in a James Bond movie, the revolving doors whirred softly, and a tall blonde dressed in a long camel hair coat walked across the lobby toward the bar where we were due to meet. She then turned to look in my direction, and her beauty froze me momentarily. I rose to my feet, and met her halfway across the room, taken aback. My carefully rehearsed Bogart style lines evaporated into thin air, and all I could manage was a "Hello, what a pleasure to finally meet you!" and kissed her on the cheek. We were both a little tongue-tied and shy, so I offered her my arm and we walked out onto the street and toward the restaurant where our dinner awaited us.

We walked easily together in step, each sensing what the other was going to do next, and arrived at Restaurant Uly, a synthesis of fusion food, Buddha bar music, fish tanks, video walls and comfortable sofas, surrounding a classic French-style restaurant with male waiters and candles. We ordered. The food was excellent, and we offered each other little tastes of our respective dishes. We stared across the table at each other. I must have had a stupid smile on my face, because I appeared to be amusing her. As I gazed fondly across the white linen at her, a thought flashed through my mind transmitted from somewhere deep in the quantum vacuum: "You are going to marry this woman—she is the one". Me? The man who vowed he would never marry again? Apparently so.

We were engaged by the end of the week, and married in September of the same year. Our tenth wedding anniversary approaches. Kismet? Divine intervention? Incredible luck? It happens, even to mere mortals such as we.

Some say luck just happens to us—others, that we make our own luck. For me, both are true, depending upon the circumstances. If I had met my wife even a few years before I did, we would not have been ready for each other—timing is very important in relationships. If I had decided to compromise on someone who was 80% of what I wanted rather than 100%, then that would not have lasted either. If I had merely restricted my search to the world I already knew, we would never have met. I had to "take a risk", even though the biggest risk would have been to settle for mediocrity in my comfort zone. Even though I do not believe in heaven in the religious sense of the word, I do believe that we can create our own version of heaven on earth if we know ourselves well enough to make the right decisions at the right time.

It was wonderful to have a partner I could share my life with, who also turned out to be a talented co-creator of our little corner of paradise. Her warm and generous personality attracted and encouraged our friends and family alike, and my children soon started loving her deeply too. I felt comfortable sharing my own feelings and opened my heart up to the world wider than I ever had before. Of course there have been the inevitable troubles in paradise, but our relationship has grown and deepened in many ways in spite of or perhaps even because of them.

The results of our early years of work together on Chateau La Tour Apollinaire have borne fruit[35], and we have delighted several thousand guests over the years. Thus was laid the foundation on which our next adventure could begin, as well as deeper appreciation of what wellbeing and thrival can look like on a very low carbon footprint.

One of the more important goals I had in moving to the south of France from London in 2003, was the desire to be able to host seminars and events in a relaxed, natural environment closer to nature than any hotel in a large city. My dream was to create a cross between a spa-type setting and a place where powerful and influential agents of change could meet to relax, refresh and reinvent themselves and the world.

We were fortunate to begin with guests such as Corbis Images (owned by Bill Gates), the French media group Hachette Filipacchi, Reporters sans Frontieres and Integral without Borders, as well as a few major local press conferences and a sports sponsorship. We were introduced to the Mayor and some of the local political personalities.

Between my plans to write a book on the great shift taking place in our world today, and a non-executive directorship in a sustainability and leadership consulting firm in London[36], I formed the intention to create a not-for-profit activity that could act as a catalyst for sustainable and social innovation. Thus was formed Renaissance2, and the Great Shift Gathering, a five-day conference hosting 100 people in a great tent in

---

[35] You are welcome to find out more and visit the website of the Chateau at: www.latourapollinaire.com.

[36] A shout here to the co-founder of Future Considerations, Paul Gibbons, change management expert and uber-coach.

the gardens of the Chateau, all exploring the future shape of our global civilization[37].

The full spectrum of the rainbow of human values that aspires to integral or second-tier consciousness and culture was on display in this event, from Achievist "Orange" business people networking, through a thousand Affiliative "Green" hugs and small group bonding sessions, through to some very serious Integral "Yellow" enquiries into some of the most pressing issues of our time. We even sensed, from time to time, a very precious space that might have been a dash of Transcendent "Turquoise", involving a wonderful peacefulness and sense of expansion into the beauty and wonder of the world and the times we are living in.

James Martin wrote in *The Meaning of the 21st Century: A Vital Blueprint for Ensuring our Future:* "Evolution on Earth has been in nature's hands. Now suddenly, it is largely in human hands . . . as we automate some of the processes of evolution the rates of change will become phenomenal. This change from nature-based evolution to human-based evolution is, by far, the largest change to occur since the first single cell life appeared . . . when it first happens on a planet, it is probably dangerous. The creatures that take evolution into their own hands have no experience in the game . . . we are an experiment in free choice."

As my dear friend and colleague Barbara Marx Hubbard put it in her introduction to my book "The Great Shift":

"This is an evolutionary driver of the highest order. The set of challenges is so complex and interactive that to the ordinary intelligent, linear mind, it seems impossible. Most people tend not to respond to this unprecedented scale of challenge."

We live in extraordinary times. The first decade of the 21st century has witnessed incredible swings between irrationally exuberant millennial optimism and end-of-the-world doom and gloom. I believe we are about to experience a calmer, more realistic phase in the next stage of the evolution of our global civilization. Why?

---

[37] Of the many incredibly talented people involved in the formation of Renaissance2, my particular thanks go out to my wife Elena Wood, Mark Wade (retired sustainability leader in Shell), Graham Boyd (innovation and leadership fundi), Matthias Lehmann, Robert Yarr, and Napier Collyns (founder of Global Business Network).

The twin ecological and economic global crises have focused the world's attention in an unprecedented manner. Major decision makers worldwide are all more or less agreed on the nature of the challenges we face as a planet, and the more complicated question is: what is the best way to go about resolving these crises in a sustainable manner?

Looking at a picture of our planet hanging like an opalescent pearl suspended in the black velvet of space usually brings tears to my eyes if I stare at it long enough. Why? For many years, the sight of spaceship earth dangling in the void inspired me, and brought home the fact that ultimately humankind shares a common destiny, whatever craziness we were experiencing down here at the time.

Now, however, when I gaze at that same image, I feel a deep sadness. What have we done? I see the rainforests being slashed and burned, the coral reefs dying, half of the planet eking out an existence on less than a few dollars a day, and species disappearing at a faster rate than ever before. And above this all, looms the tipping point for climate change, the point of no return, where runaway global warming takes hold and we end up in a world where temperatures have risen by 4-6 degrees C on average, the rainforests are burning and methane escapes uncontrollably into the atmosphere from the Siberian and Canadian tundra. This is happening right now, ten times faster than ever before in the history of our planet.

Our exploitation of the "last hours of ancient sunlight" through the extraction, pumping and burning of fossil fuels in the past few centuries has accelerated us toward a tipping point both in terms of climate change and in terms of the fact that we are the last generation that can (or should) rely on fossil fuels as our principal source of energy. I would argue that the UN should add a fifth major challenge to the four global challenges of our time above: the materialistic aspirations of the two and a half billion people on our planet who are now officially "middle class". For even half the human race to live like people in the most advanced cities on earth, we know we would need between another four to six planets of resources—and we only have one planet.

There are many great, civilizing attributes to the modern middle-class which now dominates the planet: it believes in democracy, not autocracy; it creates a society of aspiration, where we hope our children will do better than we did; it believes in merit not privilege; competition not inheritance; it applauds thrift and personal effort, and most importantly for our purposes, it begins to value nature and the environment.

We live on a planet where 2.5 billion people are living modern middle-class lifestyles, 2.5 billion are living in relative poverty in traditional societies, perhaps 300 to 400 million are classified as "rich" with modern worldviews and lifestyles, and another 300 to 400 million are "cultural creatives" and "integrals" who are world-centric in their outlook.

In shifting from unconscious to conscious design of our future world, the next steps we have to take include:

- Connecting up, aligning and supporting those who are actively shaping the great shift that is underway, so that their efforts are multiplied, leveraged and accelerated in as many ways as possible.[38]
- Creating global, trans-disciplinary research, development and innovation programs in the areas of our built and natural environments, human wellness and the knowledge and design environments, to ensure that our bodies and our world can thrive beyond the 21st century.
- Catalyzing the conscious evolution of integral, world-centric culture through innovative approaches to the development of human consciousness and potential ("goodness"), the advancement of our symbolic intelligences and narratives ("truth"), and the expression and appreciation of our civilization in all its magnificent forms ("beauty").

Integral theorist Steve McIntosh[39] believes that in order for any form of world governance to work, we need a sufficient number of people on the planet operating at a world-centric level of consciousness, which he defines as: "A worldview beyond postmodernism that recognizes the legitimacy and "evolutionary necessity" of the many different stages of development in consciousness, cultures, and individuals around the world today". When enough people begin to adopt this worldview, McIntosh sees the possibility of a "world federation" emerging whose

---

[38] This is one of the roles of **Renaissance2**, the organization of which the author is the founding President. See www.renaissance2.eu

[39] Integral Consciousness and the Future of Evolution, Continuum, 2007

objective is to "harmonize the needs of the modern and postmodern developed world with the needs of the traditional third world."

In the 21st century, the convergence of breakthrough technologies, socially beneficial ways of doing business, various forms of widespread spiritual awakening, integral philosophies and a new world-centric approach to politics, offers the human race a rare evolutionary opportunity to make a great shift toward a sustainable, thriving world civilization. What is needed is a shared, joined-up vision of the potential kinds of futures we might create together at a global level: a new "world vision".

To sum it all up, here is a simple equation: the survival and future thrival of you, me, and all of us, will be decided by two competing forces in the next decade or so:

The Bad News: The Speed of *Climate Change* together with the Rate of *Ecological, Economic & Social Decline*

The Good News: The Speed at which we design and build a *Renewable Economy* together with our *shift towards World-Centric Consciousness*

Right now the balance between the bad news and the good news is tipping toward the dark side, and only a major acceleration in our shift toward a renewable economy plus a shift toward world-centric consciousness can ensure we make it.

Given the failure of big governments, corporations and banks, big media and large non-governmental organizations to prevent economic, ecological and social meltdown, I believe that a new approach is required to ensure that we create the conditions for the continued existence of human civilization on planet earth.

This is an exceptionally complex problem, which cannot be solved with simplistic slogans, movements or obvious solutions. Yet, even if you agree only partially with my analysis of this situation, it is hard to argue with the important shifts which need to occur rapidly in our global political, economic, social and technological systems if human civilization as we know it is to survive the 21st century:

- **From *growth and size* to *development and scalability*—**our current reliance on big governments, corporations and banks, big media and large non-governmental organizations to get us out of this mess needs to shift to self-reliance and responsibility leading to activism in our families, communities, organizations and networks to create the systems we need for a sustainable future;
- **From *national and transnational* to *world-centric* systems of governance—**we need to shift our current "top-down" approach to organizing our world to include a much stronger and healthier "bottom-up" approach, creating genuine participatory democracy, while also ensuring that we enhance global standards ensuring ecological, economic and social wellbeing;
- **From a *global economy powered by carbon-based fuels* to a *glocal economy powered by renewable sources of energy, food, housing and transport*—**there is no energy shortage on earth—simply a shortage of the will and imagination required to design and deliver a renewable economy offering a better quality of life for all who wish to benefit from the cascade of innovations being implemented right now around the world;
- **From *conventional, power-driven consciousness and leadership* to *post-conventional, distributed creatorship*—**in addition to harnessing the energy of the sun, we need to harness the creative talents of the world's population by demonstrating how they can develop and apply innovation to improve their quality of life without costing the earth.

In order to catalyze these shifts, we also must accelerate the shift in human consciousness from unconscious evolution to conscious evolution. This means giving people the power, the tools and the resources to become fully conscious of their impact on the lives of other living creatures while at the same time enabling them to experience the joy of a long and satisfying life lived with purpose. I believe that together, we can do it.

Dr Robin Lincoln Wood

To Sum Up: Ten basic propositions about how our universe works, and why it matters . . . .

Though there may be a great deal of trouble in paradise, and though we are nowhere near figuring out exactly where our current efforts will lead us in resolving some of the troublemaking elements, our scientists, engineers, psychologists, politicians, designers, artists and others have certainly made a great deal of progress. If one was challenged to sum up in ten key propositions what the latest science is telling us from a practical perspective, then here is where I would begin:

1. The universe we are living in is a stunningly beautiful, three-dimensional holographic projection of an implicate order—in other words, an order hidden to our eyes and other senses, but which generates everything we see, feel and are. This implicate order is enfolded in several dimensions inside spacetime, and is described by physicists using a variety of theories ranging from quantum mechanics to superstring theory[40]. There are many wonderful, thick books you could read to try to understand all of this, but as even the physicists themselves cannot agree on exactly what lies inside the quantum, you would do better to read the next proposition before diving into such tomes.

2. Our external senses respond to what is outside of our skins, while our internal senses respond to what is inside our skin. This may seem a little obvious, but the distinction becomes important later on, as you will see. The implicate order is the source of both our interior perspectives, as well as the way in which we interact with the explicate order outside of ourselves. We project our interiors into the explicate order through our intentions, while responding to the

---

[40] As I write one of the emerging favourite theories is that the two-dimensional information field on the cosmological horizon of our Universe is generating the three-dimensional hologram we perceive as reality. Some imaginative experiments are currently underway to test if the core fabric of the universe is indeed "fuzzy", as would be expected in a holographic projection. http://www.nature.com/news/simulations-back-up-theory-that-universe-is-a-hologram-1.14328

explicate order through our senses and actions. Most of the time we do not notice this is happening, as many of us are wandering around this planet with our eyes open but as fast asleep as the captain of a Boeing 747 flying across the Pacific Ocean to Los Angeles with the autopilot on. Our intentions and creations arise via the perspectives we apply to our thoughts, dreams, emotions and the ongoing storytelling we do both alone and with others. You may even notice you are telling yourself a story as you read this, which is only natural.

3. Everything in the universe is interconnected. Most connections are weak or transient, but human existence is often shaped via strong connections that arise and persist over long periods of time through a process of entanglement. Our intuition provides us with multiple ways of sensing the stronger and more enduring interior connections with people, places and events which shape us and our lives, while life also provides us with kosmic nudges through the explicate order in which such connections evolve. You might, for example, notice that certain people showed up in your life at just the right time and place, and for just the right reason, and that nothing was ever the same again. That's what we in the philosophy business call a strong connection.

4. The evolution of life and the universe unfolds through the transformation of a practically infinite range of possibilities in the implicate order, into observable phenomena in the explicate order. Our role as conscious beings is to choose amongst the myriad possibilities which present themselves to us, bringing some to life while ignoring or challenging others. To make decisions, however, we need to actually be aware that we have a choice. If you are sleeping with your eyes open, you may miss some of the important decision points and you will be wondering forever after what happened, and why things are not working out exactly the way you had hoped. Life is designed so as to present you with the same choices disguised as different situations, over and over, until you are able to work out which is the right path to take. Then you get to move onto the next set of choices. If this all sounds rather tiring, it is, and I will pretend not to notice if you go back to sleep. Sweet dreams.

5. Consciousness is a fundamental property of our universe. As conscious beings we are the universe aware of itself, which is pretty cool as it goes. But what, you might ask, is the point of consciousness? Exactly—fine question. Well, there may never be a truly scientific answer, which is where one needs to choose a philosophical perspective to answer this question: here is mine.

6. The role of consciousness has been and continues to be to guide our evolution towards beauty, truth, justice, wisdom, responsible freedom and love. Being conscious means we are able to sense when and how to move toward these ideals in the concrete world of the explicate order, as we interpret these ideals through our own unique set of perspectives each and every moment of our short yet action-packed lives here on this increasingly warm third planet from the sun. Human development revolves around the fact that there is only one person ultimately responsible for your development: you. Yes, I said you, not the lady behind you. And as you grow in your development, you will become increasingly responsible for the development of others, and you will find that those you are closely connected to are also responsible for your development. But beware—they may not always have a label such as "coach", "leader", "guru" or "teacher" on their forehead. Indeed, they may appear to be the opposite of a responsible person or leader, and that, in the ironic kosmic scheme of things, is just the point.

7. Integral psychology tells us that the process of human development is pretty much open-ended, and theoretically without end. At this precise point in our evolution, however, while we are only in the planning stages for a manned mission to Mars and still trying to work out how to get people to stop killing each other in many parts of our planet, it is probable that most people will experience between five to seven developmental stages in their lifetime. The most advanced beings on earth may experience ten or more stages, but that takes an awful lot of carrot juice, meditation and yoga (just kidding) ;-) The development of each and every person is also neither necessarily aligned nor balanced—for example, you may be brilliant in just one developmental zone, in which case you are less likely to have the time or inclination to be brilliant or even good at

anything else. Or you may be good at a lot of things, but this often precludes you being brilliant at any one of them. It is still rare to find people who are physically, emotionally, mentally and spiritually advanced and balanced, but luckily for us, there are now enough of them to ensure we aren't all vaporized in a bright green fluorescent hydrogen cloud.

8. The healthy growth and development of people requires a healthy environment in which that development can take place. Health starts with a positive state of wellbeing, not just the absence of disease. A healthy environment is not simply an environment without challenges, for that would not stimulate development. And even more frustratingly, what looks healthy from one set of perspectives may feel positively unhealthy and claustrophobic or wildly risky and dangerous to another. It is easy to see how a well-functioning health-care system called "Mom" is necessary for the wellbeing of her child, yet at a certain point that child will need to learn to deal with dysfunctional and unhealthy "others" to grow and learn some of life's harder lessons. But there still needs to be a secure base for healthy development, however dysfunctional certain aspects of your life conditions are.

9. At the end of all the effort, fun, pain and joy of this development, we all know we are going to die here on planet earth—that part is relatively straightforward. In the explicate order, the telomeres at the end of our DNA wear out, our cell repair processes fail, our auras lose their shine, and we become ill or simply snuff out like a candle with a heart attack or a stroke—(the lucky ones go in their sleep, while astral travelling). In the timeless and infinite implicate order, however, there is no death. Whatever has existed will always continue to exist, though it will have evolved beyond all recognition through the eons. Death loses its sting, looked at from the implicate order side of things. Now here's the trick: all information in the universe is not only preserved (even inside a black hole), but is accessible if you can dial up the right frequency.

Yes, life in the explicate order may defiantly cock its negentropic nose at the second law of thermodynamics which causes all systems to run down over time. But remember that each and every moment the explicate order is unfolding from the implicate order, bubbling up chaotically from the quantum soup to be harnessed by life as light, energy, matter and information. There are many second—chance saloons spread liberally throughout the giant spinning galaxies out there, and as long as you are able to learn and develop, nothing that has ever happened to you (or even for that matter, things that did not happen to you), can be regarded as a total failure (or for that matter, a total success). The explicate order needs entropy and decay to ensure that the information and learning you've generated in your life can be returned to the implicate order. Death is simply the universe's way of telling you to stop, so that you can begin the next adventure.

10. Old fashioned ideas like faith, hope and love might sound like religious bunkum read out from a pulpit on a Sunday morning, but viewed from the perspective of evolutionary philosophy, they are simply the universe's way of creating the space we all desperately need to transcend our exterior circumstances and make room for the future. We must be able to suspend our disbelief and world-weary cynicism in order to move on in our lives. We must develop a faith in our ability as a species to co-create a thriving world through the application of evolutionary philosophies and technologies, rather than a faith in other worldly gods who somehow determine "how it is", or a dogmatic materialistic scientism that says emphatically that "Exteriors Alone Are Us".

Healing, growth, transformation and redemption are possible if we:

- Engender faith in our desire and ability to make the shifts we need, personally and collectively
- Design and build the socioeconomic operating system that gives us hope for a thriving future
- Express our love for ourselves, others and the worlds we co-create, no matter the difficulties.

The old cliché says it best: the future really lies in our hands. We can start to create "heaven on earth", beginning with our own small corners of paradise, right now. No matter what challenges the world may throw at us, and how much trouble there may be with paradise from time to time, our lives can become an ode to joy in every moment if we only choose to make it so.

Epilogue
Part One

In the few months this book has been in the hands of my publishers, one or two interesting items have come to my attention that I believe would add to your enjoyment and edification. Firstly, I promised you a newer version of the poem, "The Lake of Eternity", written by my ancestor all those centuries ago (in the late 1700's in Stellenbosch South Africa), featured in chapter 2. What you see below was written in the midst of my transition from a "psychic atheist" to a "spiritual agnostic" in the mid 1980's subsequent to my emigrating to London:

Memoirs
A Reflection on a Poem by my Ancestor Meent Borcherds Written 250 Years Ago

Mighty Universe, grant me this hour
peace
I look back across time
generations of word made flesh
the future
a giant whorl
spinning slowly
at the edge of the vortex
sinking into the lake of eternity
strands of dna
attest to your being
helicyclical spirals
of being and becoming
driving me
out,
up,
beyond
time
space

thou art, I am
one
out of stardust
order out of chaos
inheritors of earth, sky and sea
meeting place of opposites
life's simple puzzle
I will be
therefore I am

"My atheism, like that of Spinoza, is true piety toward the universe and denies only gods made men in their own image, to be servants of their human interests"

In 2014, inspired by many influences, including integral philosophy and psychology, I wrote a follow-up prayer to the universe, reflecting an approach both more scientific **and** mystical based on what I've learned in the past 25 years.

An Integral Prayer for ThriveAbility

Let us bow our heads . . . .
Oh, Conscious and Loving Universe,
Ever evolving toward greater Kosmic Beauty, Goodness
Truth, Wisdom, Caring and Compassion.
Hear my Integral Prayer . . .

1. May my being and my behaviour be of service to All Beings in All Worlds, liberating all into the suchness of this and every moment.

2. May my intentions and choices in this moment and the next be beneficial to the Flourishing of all life, everywhere, for all time.

3. May my view of the world both transcend and include all beings at all levels in a profound and irresistible integral embrace that supports each life to stretch up toward the stars.

4. May the Kosmic LifeForce grant me the wisdom
of discernment, and the joy of peace in communion
with all life wherever it may be.

5. And may I find liberation for myself and others through
the expression of my unique evolutionary self to find the
fulfilment of my purpose here on Earth in this lifetime.

6. Grant me in this moment, Oh Universe, the courage to face
and overcome my own challenges so that the suffering of
myself and others is transformed into the profound and lasting
transformation our world desires with such fervent longing.

7. May the peace, love and wisdom that surpasses all
understanding be with us always, even in the shadow
of the Valley of Death, and in our darkest hours.

8. And may the loving life force enter into all of those at war
with themselves and others, and grant them peace.

9. In the name of all that we each hold sacred and holy, may this
all come to pass in the fullness of each and every moment.

10. May we now each be filled with and carry into our day
the Radiant Joy that comes from realising ourselves
and others as we truly are, and turn our Original Face
to the warm and glowing light of our Oneness.

One Thriving World, One Glorious Taste, Forever and Ever,

Amen

Epilogue
Part Two

The Story of ThriveAbility so Far

Humans have been and are today most delicately poised as the **fastest evolving species** on the planet and now play a role as the **most sensitive transducer** of what it means to be alive and well with purpose and joy. *ThriveAbility focuses on the flourishing of all human beings so that they are able to fulfill their individual potential to the fullest and promote a positive state of wellbeing in a sustainable environment beneficial to all life.*

A few years ago I was preparing a speech on "Co-Creating a Viable Future—Vision 2050" for a conference. I sat there in my study wondering why, after half a century of major effort, environmentalists and practitioners of sustainability were still failing to close the gap between what we need for a viable future, and our current situation where we are using 1.5 planets worth of resources and heading for 3 degrees global warming. We must now deal with the challenges of accelerating climate change and the hard fact that sustainability 1.0 is not closing the gap fast enough for a viable future for mankind.

It suddenly occurred to me, in one of those "AHA!" moments, that the problem was the **equation** being used by environmentalists and practitioners of sustainability, with all its focus on impact reduction. This focus on risk reduction and optimization of the existing system fails to motivate the majority of people to take the actions needed, and to innovate in ways that deliver the breakthroughs we need. It takes a "one size fits all" approach to people, organizations and countries, rather than tailoring its approach to meet people where they are at and motivate them to think and behave in often radically new ways.

ThriveAbility offers an elegantly simple way of ensuring the survival of our species by finding new ways of integrating *sustainable breakthroughs* into *thriving lifestyles, organizations and communities*. The ultimate goal is to create a new attractor to encourage actions that enable life conditions for healthy human emergence and a thriving planet. ThriveAbility offers frameworks and insights that radically simplify many of the tough decisions we have to make on a daily basis to ensure our wellbeing and that of future generations. *In short, we learn to live lives that thrill us with their richness and lightness of footprint, while inspiring others to follow in our footsteps*

The Shift Needed

"Our current economic model is a global suicide pact.
We mined our way to growth, we burned our way to prosperity.
We believed in consumption without consequences. Those days are gone."

Ban Ki-Moon, UN Secretary-General

Most of us are now aware that our current economic system is based on an unsustainable, high stress, linear economy powered by fear, fossil fuels, materialism and a myopic focus on financial success. Maintaining this rationale will both cause irreversible damages to planet Earth and accelerate the rapid rise in social dissonances due to stress, suffering, population growth, demographic change and increasing social inequalities. Ironically, this situation is caused by the very success of modernism in creating material abundance through conventional science, technology and corporate business systems.

Contemporary society has reached a crossroads: We have realized that we are on a highly unsustainable path and have to move towards a fundamentally new one. Not only have we reached the limits to growth in the linear economic paradigm, but the question of how to measure economic success simultaneously raises the highly philosophical question of everyone's 'meaning of life'. What do we live and work for? Is it for the continuous improvement of black bottom lines or is it for the creation of a thriving society? Is material wealth the ultimate indicator for societal prosperity or rather aspects like health, happiness and education? New indices like the Happy Planet Index or Michael Porter's Social Progress Index have emerged during the past years to expose the limits of currently predominant success measures and draw our attention to the fundamental idea of ThriveAbility.

We are all now "Managing Directors" of Planet Earth. Some of the signs of our current poor performance in this role include the tremendous rise in greenhouse gases in the atmosphere together with the associated rise in global temperatures and sea levels, the large global increases in soil erosion caused by land clearing and soil tillage for agriculture, and the massive extinctions of species caused by the widespread destruction of natural habitats.

In order to shift sustainability for the breakthroughs needed, we need to put it into a broader context, by combining it with innovation, design, big data and the latest insights into how human consciousness and motivation work. This leads to a more holistic approach and understanding of transformation, starting with the individual and empowering the many. This combined understanding also embraces concepts like social entrepreneurship, flourishing, the circular economy, integral psychology; concepts that all come from certain angles, but also leave out others. That is ThriveAbility's promise—to integrate, and build on the power of existing approaches through a radical simplification—or 'simplexity'—the simplicity on the other side of complexity when viewed through a powerful unifying lens.

The vision of ThriveAbility is to serve as a waymaker for the next economic paradigm, building on the idea of a Conscious Economy. The aim is to integrate existing approaches into a new worldview that supports mindsets, models, methods and practices that have the potential to facilitate a systemic transition towards sustainable thrival. ThriveAbility is a fractal concept that ranges across different scales of time and complexity—it is applicable from the level of individuals, families and communities right up to organizational, national and global levels.

ThriveAbility has been designed from an integral perspective, to ensure that we cover each and every major aspect of the challenges and opportunities involved, ensuring that the outcomes produced are themselves thriveable and sustainable. In diagram 1 below the four "quadrants" of the ThriveAbility Template are shown:

- The two interior quadrants, being "Psycho" (psychological/spiritual) and Cultural (norms and values);
- The two exterior quadrants, being "Bio" (biological/behavioural) and "Socio" (socioeconomic systems).

The Trouble with Paradise

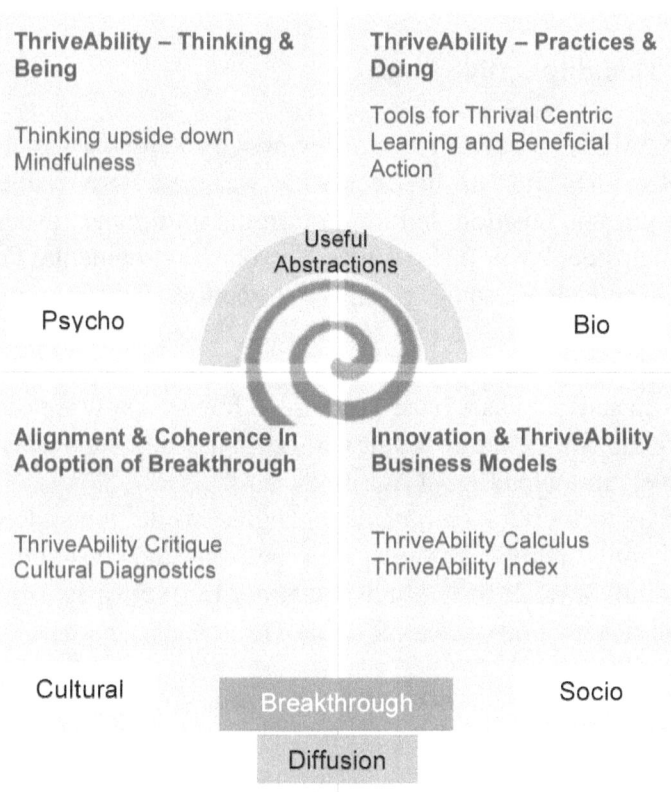

Diagram 1: The Four "Quadrants" of the ThriveAbility Template[41]

---

[41] Each of the four quadrants are described in greater detail below.

1. The "Psycho" (Psychological/Spiritual) Quadrant—ThriveAbility Thinking & Being

o **Thinking Upside Down**

The "AHA!" moment I mentioned above about why sustainability was not closing the gap between what we need for a viable future and our current situation, led me to turn the traditional sustainability equation upside down. Instead of putting our Environmental Footprint on the top and our economic growth on the bottom, I put Human Thrival (or "Flourishing") on the top, and divided this by our Environmental Footprint. Rather than seeking footprint reduction only, while viewing our human activities as a "cost", I chose to focus on how we can create better lives, with smaller footprints, by being innovative and being collectively more intelligent and wise.

Thrival thus offers a motivational engine, while Impact Reduction generally only motivates "deep green" activists and consumers, while offering only a weak form of advantage for most businesses pursuing CSR and sustainability initiatives. The ThriveAbility model:

- offers a comparative integration of what we currently know, which then provides us with a base from which we can each begin to make better decisions immediately while accelerating our learning process with new and better information as we begin implementation, avoiding both analysis paralysis and suboptimal "shoot-from-the hip" decisions. places in perspective
- both the human and environmental trade-offs and synergies available in any situation, at any level of scale, from the personal to the global and all layers of complexity in between.

This then engenders a new way of thinking and being: "ThriveAbility Thinking", which goes beyond the currently most advanced thinking frameworks available, namely integral and wellbeing psychology and advanced sustainability and innovation metrics embedded in dozens of incompatible frameworks.

Most importantly, ThriveAbility Thinking enables us to get into our strategic psychological helicopter, transcend the situation and transform the outcomes possible in making often politically fraught, hard decisions.

The knowledge base for making better decisions through ThriveAbility Thinking is a continually expanding, global public good rooted in the communities of practice which comprise the ThriveAbility Consortium, a not-for-profit guardian of the principles and practices of Embedding ThriveAbility.

To Embed ThriveAbility in your life, family, community, organisation or nation, you will need to make ThriveAbility Thinking a habit, which is always able to answer the question:

"In the current situation, what actions will yield the best outcomes for the greatest number while promoting the capacity of the group responsible for this decision to successfully implement this decision across a range of scenarios?"

o **Mindfulness and Mental Models**

*Achieving sustainability is as much a challenge of **changing our mental models**, as it is a problem of **making flows of materials, energy and other resources eco-efficient**. Being mindful is the key to accelerated learning.*

Concentrated attention, reflective awareness, presence and reflective learning are all terms used to describe "mindfulness". From a ThriveAbility perspective, mindfulness is much more than a nonjudgmental present-centered form of awareness of one's experiences—it also brings to bear an "overview" perspective of one's choices and options in the moment, based on an intuitive understanding of how one's choices enhance the thrival of all life. As we grow and mature, so too do our values embrace ever wider spans of beings and worlds.

The consequences of such mindfulness result in the ability to make decisions and plan actions that will have the tendency to not only center and deepen one's own present state in an experience of flourishing and wellbeing, but will also ensure that others and all living systems benefit from this state. there is a direct connection between the practice of mindfulness and the cultivation of morality

Learning in such a mindful state is also accelerated, enabling us to shift our "mental models" into more productive and beneficial patterns. The old cliché is that "Change = Learning and Learning = Change".

Mindful Learning = Beneficial Change

2. The "Bio" Quadrant (Biological/Behavioural)—ThriveAbility Practices & Doing

o   Tools for Thrival Centric Learning and Beneficial Change

Living well in ways that do not cost the earth is going to become a core skill for those of us transitioning from an unsustainable, linear economy, to a thriving, conscious economy. The big question is how to do this wisely and intelligently, as well as creatively and joyfully, rather than regressing back to more primitive and less satisfying lifestyles. Moving toward a Conscious Economy, means, amongst many other things, learning new skills and developing ourselves and others so that through our daily practices we are:

- transcending the materialism of modernity (growth for growth's sake), and getting more out of less while living a better quality of life;
- making choices based on a deeper understanding of the ingredients of thrival thereby enhancing our own wellbeing and that of our families, communities, cities, organizations and nations;
- embracing the full spectrum of human potential (developing oneself and others) by building on our own strengths and talents as well as those of others;
- enjoying the richness of diverse cultures, sharing common perspectives and reframing who we are within a global community (celebrating diversity);
- working for a planet that is increasingly alive, conscious & reaching for the stars (enjoying higher levels of consciousness and aliveness).

The tools and methods available for such practices are available in literally every corner of the world if we know where to look, and the ThriveAbility Consortium can recommend a dozen good places to start if you are interested. Just one example—Integral Life Practice (or "ILP"), blends physical, emotional, mental and spiritual practices that can be done on a daily basis in a busy schedule. Other more mundane but important practices such as recycling, buying sustainable products, eating healthy, local food, planting gardens and investing in sustainable businesses that promote ThriveAbility are also simple ways in which to begin to make a difference. Making all of this as fun and enjoyable as possible, along with healthy doses of music, art, dance and thriveable entertainment, is important too.

3. The Cultural Quadrant (Norms and Values)— Alignment and Coherence in Adoption of Breakthroughs

o **The ThriveAbility Critique**

One of the key differences between ThriveAbility and other approaches to ensuring a thriving human civilization on a thriving planet in the 21st century, is the fact that practitioners of ThriveAbility are not only self-aware but also conscious of the way they are thinking about specific issues moment to moment. In other words, ThriveAbility practitioners are aware of their own worldview in operation, and able to shift worldviews to find more appropriate ways of perceiving and thinking about a situation or challenge.

Depending upon one's sources, there are at least a few hundred million people alive today who are capable of such perspective shifting, and the number is growing daily. Such a shift in perspectives can yield a much more effective response to situations, challenges and opportunities, and ins some cases, yield an outcome between ten to a hundred times more effective and ThriveAble. Some of the ingredients of the ThriveAbility Critique include:

a. **Perspective Awareness**—What is the thinking/worldview driving the way you are currently approaching a situation, challenge or opportunity?
b. **Options for Change**—what are the different ways in which one might approach changing the situation and moving toward a more ThriveAble outcome? (Including the eight possible kinds of change response inherent in the Gravesian change model)
c. **Creating Conditions for ThriveAble Change**—what would be the most appropriate methods, tools and technologies to support profound change in this situation? How can different worldviews

and interests be "streamed" so that potential conflicts can be effectively managed? What is needed to ensure that the changes "stick", and that they do not fizzle out after the initial stimulus and enthusiasm wear off?

d. **Designing a Change/Transformation Process**—the ThriveAbility Critique is a useful first step in thinking through how a variety of visions can be translated into a design process to map out a change/transformation process. This needs to be done in conjunction with the tools for cultural diagnostics and alignment below.

o **Cultural Diagnostics and Alignment**

How can apparent conflicts and blockages in families, communities, organizations and nations be resolved during the change process? How can different visions of the future be aligned to create the synergies required in social systems at all scales? A variety of proven methods are available to help us move toward more ThriveAble outcomes:

a. **The Personal, Organizational and Social Alignment Wheels**—Over the past quarter of a century a variety of different action research programs have resulted in the development, testing and effective application of cultural alignment wheels at three different scales. These cultural alignment wheels have been proven to result in better strategic performance and more ThriveAble outcomes.

b. **Cultural Fit Factor**—using approaches developed from Gravesian roots with a thorough scaffolding of integral psychology, it is possible to assess where individuals and teams "fit" into a culture, at any scale. Although developed and tested mainly at an organizational level, the basic principles on which these test are based enables them to be applied to much larger groups as a tool for the analysis of social cultural fields, and the resonance or dissonance being experienced by different individuals and groups within such social fields.

c. **ThriveAbiity Inquiries**—much like appreciative enquiry before it, a ThriveAbiity Inquiry not only examines what is already working within a social system, it also identifies the blockages to change and transformation. When used in conjunction with alignment wheels and cultural fit tools, a ThriveAbiity Inquiry is a creative, dynamic process that both inspires the participants and also releases tremendous amounts of blocked energy that can be put to purposeful use in building the next stage of the system in focus.

4. The "Socio" Quadrant (socioeconomic systems)— Innovation and ThriveAble Business Models

o **Breakthrough Innovations**

Innovation is US—We humans are alive today because our core strength is our ability to adapt and innovate to a constantly changing and often hostile world. Innovation is also evolution in action. Every stage of human development comprises a coherent and integrated way of making a living, exchanging and distributing value and dealing with shocks and surprises. The rate at which innovations are adopted is driven by the evolution of personal, cultural and social norms and habits, in the cultural quadrant we examined above.

At each level of human development, there are specific **Limits to Growth** and predictable **Crises** that need to be resolved through combinations of the tens of thousands of innovations that have been recorded since the dawn of humankind. **From Fire to Freud**, we overcome limits to knowledge and growth. Disruptive innovation drives breakthroughs based on removing artificial, often self-imposed constraints rather than more fundamental limits.

Each of these innovations is based on an **Original Idea** (UL), supported and scaled within a set of **Values and Needs** (LL) that create new opportunities and options for our species at each stage of development. Once they are adopted by the mainstream, such innovations are taken for granted and become "baked into" our civilization as an enabling layer of infrastructure, tools, habits and norms.

This in turn creates the crucibles in which new innovations can emerge to meet new life conditions. For example, the innovation of ThriveAbility builds on previous innovations in design, strategy, leadership, sustainability, integral philosophy, psychology and practices, and the art and science of disruptive innovation itself.

o **ThriveAble Business Modelling**

Ideally, a "business model" is a way of conceptualising the design of a set of virtuous circles between human needs and the capabilities/possibilities of a supplier or network of suppliers, mediated through a set of fair exchanges.

Socioeconomic innovations can be scientific, technological, product, process or business model based. We are all familiar with the way in which household products are continuously being improved, and increasingly often with sustainability in mind. For example, most manufacturers are now removing excess sugar, toxic substances and other harmful ingredients from their products, while also developing more eco-friendly packaging—they see this both as a moral necessity as well as a form of competitive advantage.

Equally, behind the scenes, many supply chains in many major industries, from food and beverages to retailing to energy to telecommunications, consumer electronics and transport, as well as building supplies, construction and engineering, are being revamped through process innovations to be more eco and energy efficient. Even the fashion industry is slowly becoming more eco-friendly, and pioneers in sustainable innovation from Interface Carpets to Puma and Nike sportswear and Unilever and P&G in foods and soaps are making major strides in reducing their footprints.

Such product and process innovations are an excellent starting point for the long journey ahead to attain what many are now calling "net positive impact" organizations. The opportunity for ThriveAbility practitioners is to accelerate these positive developments, especially through innovations at the level of business models and business ecosystems, where some truly major breakthroughs are now possible, in combination with the scientific and technological breakthroughs popping up everywhere in labs, social enterprises and "frugal innovators" in developing countries.

In particular, the cradle-to-cradle/circular economy and sharing economy business model approaches will all result in radical changes to the way we design, make, consumer and service products and services in the future, and offer tremendous opportunities for ThriveAble Business Modelling Breakthroughs.

ThriveAble business modelling can benefit from the use of software that enables innovation teams to experiment with the different components of a business model, and examine how the "wiring diagram" in a specific business, foundation or NGO can be modified in order to produce a new design that can be tested to see whether it is more ThriveAble than the previous version.

At the level of an entire business ecosystem, such ThriveAble Business Modelling can enable major synergies between different players across different industries to learn "new tricks" from each other, and to joint venture and partner to make the entire business ecosystem more ThriveAble.

o **Applying ThriveAbility Calculus to Measure ThriveAble Outcomes**

One of the greatest challenges faced by organizations of all kinds today in the rush toward sustainability and thrival, is the lack of holistic measuring systems enabling us to compare "apples with apples". Every organization, and every industry, tends to have its own set of protocols, benchmarks and measurement systems, which mean that attempts to compare progress toward ThriveAbility on a level playing field are difficult if not impossible.

From a consumer perspective, there are some apps that offer a degree of comparability between products on supermarket shelves (mainly in North America, such as "GoodGuide"), but at an organizational level the latest corporate reporting systems such as the Global Reporting Initiative are still struggling to get organizations to deal with material issues in their reporting systems, rather than simply choosing to measure what they can where the light is brightest. This is known as the "context gap", as while we have global metrics on critical ecosystem issues such as atmospheric greenhouse gases, ocean acidification, poverty, water, food supplies and so on, the organizational reporting systems are generally disconnected from such big picture metrics.

"ThriveAbility Calculus" has been developed to act as a starting point to resolve some of these reporting issues, and is designed to work in conjunction with the ThriveAbility Index currently in its early stages of development. ThriveAbility Calculus enables us to assign a ThriveAbility Factor to any organization or social system at a point in time, and then to enable us to track changes in that ThriveAbility Factor over time. The change between one measurement point and the next gives us the ThriveAbility Index of that organization or system at that point in time.

o **The ThriveAbility Index and Dashboard**

To apply ThriveAbility at all scales in any social system, from an individual to a planet, we have developed a Dashboard that illustrates the essence and five key ingredients of ThriveAbility Thinking and ThriveAbility Calculus. The application of the Dashboard is also possible on different levels, making it immediately useful to any user. The Dashboard shows that ThriveAbility has the potential to serve as 'the glue' between the hard, technical systems and the soft, human systems, when both work together synergetically.

ThriveAbility is a fractal concept that ranges across different scales of time and complexity—it is applicable from the level of individuals, families and communities right up to organizational, national and global levels.

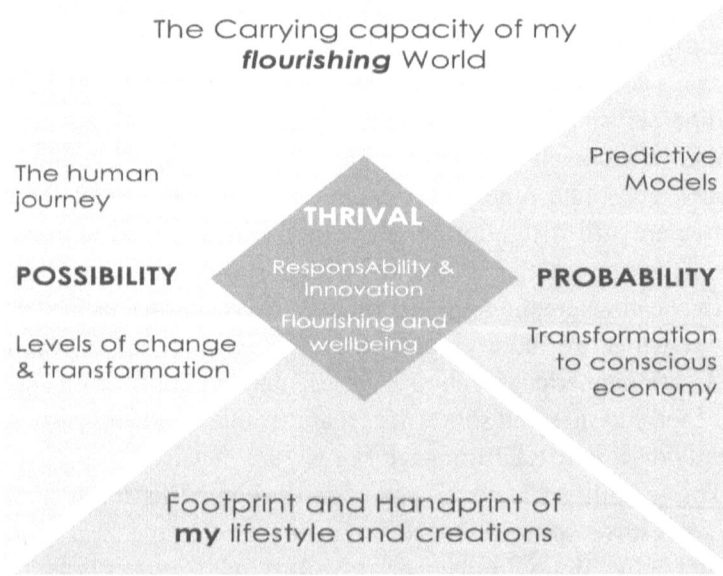

Diagram 2: The ThriveAbility Dasboard—5 Key Elements

In the table below you will find a short description of each of the 5 key elements. What follows is a brief overview of how these elements interact in practice:

- *Vertical Dimension—The ThriveAbility Equation.*— Items 1 (*Carrying Capacity*) and 2 (*Lifestyles and Creations*) are the two key ingredients for the ThriveAbility Equation. To survive and thrive into the 22nd century, we humans will need to develop lifestyles and creations which have a smaller footprint than the carrying capacity of the parts of our biosphere and social systems they rely on for their sustenance and inputs.
- *Horizontal Dimension—Transformation to a Conscious Economy and Society*—Items 3 (*Human Development*) and 4 (*Predictive Modelling for Transformation to a Conscious Economy*) bring together two of the most powerful ingredients we know of to accelerate our shift toward a conscious, thriving economy: human potential and collective intelligence. Applying integral psychological models and methods to accelerate and improve the quality of human development, we are able to connect people up in synergistic ways that enable them to harness the latest data, models and insights into the pathways available to them to shift towards thriving lifestyles, organisations and a conscious economy based on ThriveAbility.
- *Center—Thrival: ResponsAbility and Innovation*—Based on the first principles of ThriveAbility that "ThriveAbility starts with me", the central item 5 emphasizes that we each need to become role models of ThriveAbility Thinking and Doing, to ensure that the fantastic possibilities that exist in and around us are converted into probabilities that ensure our survival and thrival. Given that the next decades of the 21st century are likely to be full of shocks and surprises due to climate change and other stressors inherent on a planet of 9 billion people trying to make a living by 2050, the Ability to Respond to both the challenges and opportunities is the key message of the word "ResponsAbility".

Key Ingredient	Description
1. Carrying Capacity	The *carrying capacity* stands for the total sum of resources and flows available to support life in abundance, without compromising the Earth's ability to regenerate and recover from damages. This is the fundamental boundary condition for all actions. For life to survive and thrive, the carrying capacity has to be greater than our lifestyle and creations.
2. Lifestyles & Creations	The *lifestyles and creations* we bring into the world all depend on the carrying capacity of the planet for their aliveness and sustainability. Today humanity is already exceeding planetary boundaries, so we need to reinvent our way of living and accelerate sustainable innovation to create breakthroughs everywhere.
3. Human Journey	The *human journey* refers to our evolution as a species, and the way in which we evolve through several different stages to become world-centric adults who care about our planet. Each stage of development has its own "hot" and "cold" buttons and bottom lines, which can help motivate ThriveAbility Thinking.
4. Predictive Modeling	Ingredient four is *predictive modeling of transformation to a conscious economy*. This refers to new methods and tools that are developed through an integration of different models and frameworks to provide hands-on support to put the ThriveAbility ideas into practice.

5. The Thrival Factor- ResponsAbility & Innovation	The *Thrival Factor* is the emergent indicator that is nurtured by all other ThriveAbility ingredients. It provides a simple measure of which pathways to thrival are likely to be most robust and which have most potential for catalyzing flourishing and wellbeing. A high societal Thrival Factor will automatically lead to the development of systemic capabilities to respond and adapt to a changing and increasing complex environment.

Modeling ThriveAbility—Where do we Begin?

In order to develop a ThriveAbility Index that provides a consistent basis for measurement across all sectors of the economy, across all nations and organizations, the first step is to build a ThriveAbility Model with clear inputs, outputs and internal dynamical equations that build on what is already know about the causal linkages between the key factors involved. Many models of the global economy and biosphere already exist, from the original Club of Rome model based on Jay Forrester's Systems Dynamics modeling, to scenario models developed by the Tellus Institute, Global Business Network and many others. The basic dynamics of most of these models are very similar:

1. **Population Growth Drives Consumption**—Increasing population (driven by better medical treatment and healthier living conditions), motivated by a desire for wealth, drives economic activity and resource consumption;
2. **Consumption Drives Pollution**—Resource consumption gives rise to waste products and pollution, which are also an important source of disease;
3. **Scarce Resources Drive War & Famine, Reducing Population**—Competition for scarce resources leads to famine and war, which together with pollution, drives up the death rate. Disease, famine and war have kept the human population small for millennia.
4. **Non-Renewable Industrial Processes Drive Population Growth**—Industrial scale manufacturing, construction and agriculture (driven

by breakthroughs in knowledge and technology), began to generate surpluses which enabled improved healthcare, sewage and sanitation and the development of drugs and antibiotics, increasing the birth rate and reducing the death rate.
5. **Educating Women Reduces Population**—Current evidence demonstrates that the quickest way to reduce the birth rate is to educate women and provide them with better health care so that they need to give birth to fewer children.
6. **The Earth Regulates its Own Temperature**—The earth's atmosphere is held in a dynamic, stable non-equilibrium state because its geology, chemistry, weather and life form a single adaptive, self-organizing system which both maintains and is maintained by life.
7. **Rising Greenhouse Gases Increase Storms, Floods and Famines**—Unfortunately, current levels of pollution, including greenhouse gas emissions such as carbon dioxide and methane, have begun to destabilize the natural carbon sequestration process that converts atmospheric carbon dioxide into limestone and chalk seabed rocks through the carbon cycle, resulting in a rapid rise in greenhouse gas levels and global temperatures.
8. **Storms, Floods and Famines Drive Greater Resource Consumption & Reduce Population**—The additional heat energy trapped in the atmosphere and oceans by excess greenhouse gases is dissipated by storms, droughts and ice cap/glacier melting causing increased flooding and destruction of habitats requiring even more resources to be consumed to counter such storms, floods and famines.
9. **Deforestation Adds More Carbon Dioxide to the Atmosphere than All Cars and Trucks**—while deforestation is slowing, it currently adds 15% of the annual increase in carbon dioxide in the atmosphere. It also contributes to the sixth mass extinction of species on land and in the oceans.
10. **Renewable Technologies & Resources Reduce Pollution & Greenhouse Gases**—renewable energy, habitats and industrial processes reduce unsustainable consumption and the production of pollution and greenhouse gases. There is, however, a time lag in this reduction as carbon dioxide remains in the atmosphere for fifty or more years, and methane for an average of twelve years.

The challenges and solutions one can draw from all the scientific evidence and models as summarized in the ten points above are wide-ranging, and can now be stated with very high levels of confidence thanks to two decades of extremely thorough research and data from all parts of the world.

The climate change-driven ecological destruction that we are witnessing today—immeasurable loss of human life, plant and animal species caused by natural disasters such as floods, droughts, wildfires and heat waves, the disappearance of vast snow caps, glaciers and almost half of the Arctic—is the result of a mere 0.8°C rise in average temperature since 1800. We can only imagine what a further 1.2 to 3°C rise before 2100 will mean for the Earth's already vulnerable ecosystems and at-climate-risk communities. Hundreds of millions of people are already suffering the effects of climate change, particularly in some of the world's poorest areas such as Africa, India, Central and South America and across the Asia pacific region, where Australia is being hard hit. Some of the absolute top priorities from a policy perspective for governments, businesses, NGO's and charities include:

a. **Renewable Energy**—Increasing fossil fuel usage to meet the world's rising energy needs will lead to an increase of 184% in greenhouse gas emissions by 2030[42] based on current policy projections, leading to a global temperature increase of between 3 and 4 degrees centigrade by 2100. Limiting the global temperature increase to 2 degrees would be possible if we invested heavily in renewable energies such as solar, wind, biomass and biofuels as fast as possible, while reducing the consumption of fossil fuels as quickly as possible.

b. **Resilient Habitats**—More than half of the world's population live in cities, and that is forecast to rise to 70% by 2050. Designing, building and retrofitting cities and habitats to consume less energy, provide more green spaces, enhance human flourishing using renewable energies, technologies and processes wherever possible for construction, transport,

---

[42] From 37 Gt Co2eq in 1990 to 68 Gt Co2eq in 2030. http://climateactiontracker.org/assets/publications/publications/CAT_Trend_Report.pdf

offices and housing is therefore critical. Supplies of food and water also need to be sourced renewably with the elimination of all pollution and waste by 2050, a goal the world's top 200 companies have already committed to as part of the World Business Council for Sustainable Development[43].

c. **Enlightened Enterprise**—The largest 2000 companies in the world account for 53% of global economic economic output[44], and probably an even greater percentage of pollution and greenhouse gases (considering that a few hundred of them are fossil fuel producers or mega-mining companies). The Global Reporting Initiative[45] has just over 4 000 members, including most of the world's 2 000 largest companies. Their goal is to reduce their negative footprint, which is a start, but the overall data set out in A and B above indicates that a large majority of these companies are making very little if any real progress in their overall impact reduction.

---

[43] See the WBCSD Report on Vision 2050- http://www.wbcsd.org/vision2050.aspx

[44] $38 trillion in revenues divided by Global Gross Domestic Product of $72 trillion in 2013.

[45] The Global Reporting Initiative (GRI) is a non-profit organization that promotes economic sustainability. It produces one of the world's most prevalent standards for sustainability reporting—also known as ecological footprint reporting, environmental social governance (ESG) reporting, triple bottom line (TBL) reporting, and corporate social responsibility (CSR) reporting. GRI seeks to make sustainability reporting by all organizations as routine as, and comparable to, financial reporting.

A sustainability report is an organizational report that gives information about economic, environmental, social and governance performance. GRI Guidelines are regarded to be widely used. More than 4,000 organizations from 60 countries use the Guidelines to produce their sustainability reports. (View the world's reporters at the GRI Sustainability Disclosure Database.) GRI Guidelines apply to corporate businesses, public agencies, smaller enterprises, NGOs, industry groups and others. For municipal governments, they have generally been subsumed by similar guidelines from the UN ICLEI.

This is due to many reasons, one the most important being what we call the "Context Gap"—meaning that the sustainability goals companies are reporting on through GRI are self-referencing, and often not very transparent, meaning that companies can suit themselves as to what they focus on, rather than be guided by the larger scale goals needed to make the changes needed to avert widespread disaster in time.

The World Business Council for Sustainable Development's[46] Vision 2050 program sets out some "must haves" including:

- Incorporating the costs of externalities, starting with carbon, ecosystem services and water, into the structure of the marketplace;
- Doubling agricultural output without increasing the amount of land or water used;
- Halting deforestation and increasing yields from planted forests;
- Halving carbon emissions worldwide (based on 2005 levels) by 2050 through a shift to low-carbon energy systems;
- Improved demand-side energy efficiency, and providing universal access to low-carbon mobility.

The Five Elements of ThriveAbility—The Basis for the ThriveAbility Model

Following the brief definitions given of the five main ingredients on the ThriveAbility dashboard given above, here follows a much more detailed explanation of each of the elements and how they cumulatively work together to create a system that enhances human flourishing with the smallest possible environmental footprint possible given a baseline state of technological and social development appropriate to the human activity system being assessed.

---

[46] See the WBCSD Report on Vision 2050- http://www.wbcsd.org/vision2050.aspx Vision 2050, with its best-case scenario for sustainability and pathways for reaching it, is a tool for thought leadership and a platform for beginning the dialogue that must take place to navigate the challenging years to come.

Thrival, by its very nature, means very different things to different people. Given the diversity of the human species geographically, culturally, linguistically and developmentally, any definition of thrival is likely to only cover the "basics" of what it means to live a good life that does not cost the earth.

A traditional list of "basic human needs" starts with food (including water), shelter and clothing. Many modern lists emphasize the minimum level of consumption of 'basic needs' of not just food, water, clothing and shelter, but also sanitation, education, and healthcare. Nobel prize winning development economist Amartya Sen argues for a "Capabilities Approach" to welfare economics and development. He includes five components in assessing capability:

- The importance of real freedoms in the assessment of a person's advantage
- Individual differences in the ability to transform resources into valuable activities
- The multi-variate nature of activities giving rise to happiness
- A balance of materialistic and non-materialistic factors in evaluating human welfare
- Concern for the distribution of opportunities within society

Subsequently, Sen has helped to make the capabilities approach predominant as a paradigm for policy debate in human development where it inspired the creation of the UN's Human Development Index (a popular measure of human development, capturing capabilities in health, education, and income). In addition, the approach has been operationalized with a high income country focus by Paul Anand and colleagues. Furthermore, since the creation of the Human Development and Capability Association in the early 2000s, the approach has been much discussed by political theorists, philosophers and a range of social sciences, including those with a particular interest in human health.

In the most basic sense, functionings consist of "beings and doings". As a result, living may be seen as a set of interrelated functionings. Essentially, functionings are the states and activities constitutive of a person's being. Examples of functionings can vary from elementary things, such as being healthy, having a good job, and being safe, to more complex states, such as being happy, having self-respect, and being calm.

Moreover, Sen contends that functionings are crucial to an adequate understanding of the capability approach; capability is conceptualized as a reflection of the freedom to achieve valuable functionings[47].

Capabilities and ResponsAbilities—The Center of the ThriveAbility Dashboard

Capabilities are the alternative combinations of functionings a person is feasibly able to achieve. Formulations of capability have two parts: functionings and opportunity freedom—the substantive freedom to pursue different functioning combinations. Ultimately, capabilities denote a person's opportunity and ability to generate valuable outcomes, taking into account relevant personal characteristics and external factors. The important part of this definition is the "freedom to achieve"; the reason being, if freedom had only instrumental value—valuable as a means to achieve an end—and no intrinsic value—valuable in and of itself—to a person's well being, then the value of the capability set as a whole would simply be defined by the value of a person's actual combination of functionings.

Such a definition would fail to acknowledge the entirety of what a person is capable of being and doing and their resulting current state due to the nature of the options available to them. Consequently, the capability set outlined by this approach is not merely concerned with achievements; rather, freedom of choice, in and of itself, is of direct importance to a person's quality of life. Take the example of fasting as a functioning; there is an important difference between fasting and

---

[47] In other words, functionings are the subjects of the capabilities referred to in the approach: what we are capable, want to be capable, or should be capable to be and/or do. Therefore, a person's chosen combination of functionings, what they are and do, is part of their overall capability set—the functionings they were able to do. Yet, functionings can also be conceptualized in a way that signifies an individual's capabilities. Eating, starving, and fasting would all be considered functionings, but the functioning of fasting differs significantly from that of starving because fasting, unlike starving, involves a choice and is understood as choosing to starve despite the presence of other options. Consequently, an understanding of what constitutes functionings is inherently tied together with an understanding of capabilities, as defined by this approach.

starving because, in examining a starving person's achieved well being, it is critical to consider whether the individual is personally choosing not to eat or whether the person cannot eat because they lack the means to acquire an adequate amount of food.

In this example, therefore, the functioning is starving but the capability to obtain an adequate amount of food is the key element to be considered in evaluating wellbeing between individuals in the two states. In sum, choosing a lifestyle is not exactly the same as having that lifestyle no matter how chosen, and a person's well being does depend on how that lifestyle came to be. For this reason, while the combination of a person's functionings represents their actual achievements, their capability set represents their opportunity freedom—their freedom to choose between alternative functioning combinations.[48]

---

[48] An extension of the capabilities approach has recently been published (2013) in Freedom, Responsibility and Economics of the Person. In this book the authors attempt to explore the interconnected concepts of the person, of responsibility and of freedom in economics and also in moral philosophy and politics. They attempt to reconcile the rationality of the individual and the morality of the person. This book presents a methodological reflexion (phenomenology versus Kantian thought) with the aim of re-humanizing the person (through actions, but also through the values and norms that lead to a set of corresponding rights and obligations that have to ordered). Freedom, Responsibility and Economics of the Person is an extension—albeit in a rather critical form—of the capabilities approach; in particular the book discusses the concept of freedom (this does not refer back to a rationality of choice as claimed by Sen's defenders). Sen's capability approach accepts freedom but as a prisoner of a purely functional freedom. Such a concept does not take into consideration the capacity of people to apply moral constraints to themselves (i.e., responsibility).

The Thrival Factor and ResponsAbility—At the Heart of the ThriveAbility Dashboard

"Civilization is in a critical state and mankind is at an evolutionary crossroads. On one hand, problems and conflicts have arisen which are global in scale and have brought society to a condition of escalating planetary crises. On the other hand, humankind's potentials for creative change, fulfillment, and benevolent control of our environment have never been greater."

Edgar Mitchell, Astronaut, Founder of the Institute of Noetic Sciences

The *Thrival Factor* is the emergent indicator that is nurtured by all other ThriveAbility ingredients. It provides a simple measure of which pathways to thrival are likely to be most robust and which have most potential for catalyzing flourishing and wellbeing. A high societal Thrival Factor will automatically lead to the development of systemic capabilities to respond and adapt to a changing and increasing complex environment.

The role of "ResponsAbility" at the heart of the ThriveAbility dashboard serves several purposes:

- The word "Ability" is used in both "ThriveAbility" and "ResponsAbility" as a deliberate reference to the idea of "Capability" as used by Sen and others. In other words, we believe that one of the most important features of empowering human beings in their own development is the recognition that the more often they make the choices that shape their own future, the more empowered and resilient their ability to create those futures will be.
- We also recognize that at different levels of human development (psychologically as well as materially), the concept of Responsibility and the capabilities implied in ResponsAbility will be different—especially in terms of the increasing span and scale of the systems for which higher levels of development can be responsible.
- In addition, the increasing influence of the sciences of human flourishing demonstrate that the materialistic western ways of

assessing "needs" and defining "poverty" and "development", completely leave out some of the most important determinants of human flourishing, from positive psychology and rituals, to community engagement to rich and robust relationships to meaningful roles and work and to non-material accomplishments especially in the cultural and artistic spheres. In other words, what Maslow defined as "self-actualization", is readily accessible in many different ways at many different levels of development.

The Human Journey—The Source of New Realms of Possibility

The *human journey* refers to our evolution as a species, and the way in which we evolve through several different stages to become world-centric adults who care about our planet. Each stage of development has its own "hot" and "cold" buttons and bottom lines, which can help motivate ThriveAbility Thinking. ThriveAbility draws upon the work of hundreds of complementary schools of psychology in defining a predictable set of developmental levels through which every human being passes on their way to adulthood, including more advanced stages of mature adult development that have only recently become the focus of detailed scientific research.

Both evolutionary and revolutionary processes drive the trajectory of human development. Evolution is slowly becoming a conscious collaborative process, driven by our hearts and minds, rather than a blind, unconscious and competitive struggle driven by unseen forces out of our control. The human species is maturing from a troublesome teenager to a more awake and progressive twenty-something, though our progress is measured in centuries rather than years.

The revolutionary aspect of our development lies in our ability to transform ourselves and our systems and technologies, making predictions into the future difficult if not impossible in most areas of life. Whether these transformations are political (from the fall of the Berlin Wall to the Arab Spring), technical (from a million songs on your iPod to cures for most major diseases), or social (think of the rise of social media and virtual communities in less than a decade), various aspects of our lives are undergoing radical transformations every decade. Imagine we find solutions to overpopulation (the global birth rate is already falling below UN estimates), and global warming/

The Trouble with Paradise

climate change in the next decade? What is hard to believe right now can become commonplace in a decade—that is the lesson of the last century.

So it is that we find the messy process of evolution delivering greater complexity and consciousness over the longer term, while being punctuated by radical breakthroughs or breakdowns from time to time. Within the general trend of "UP", we find both major progressions and major regressions, with the latter often acting as fuel for the former, embodied in the formula: "What does not grow, decays". In the human body we find the rate of cell growth and cell death sufficiently evenly balanced so that we do not notice that our entire body is renewed on average every seven years. In stable societies and civilizations, we also find a similar ratio of growth and decay in the population and the artefacts and infrastructure they live in.

What drives the evolution of human beings is the interplay between the growth and development of individuals, and their environment. Historically our environment dictated our life conditions, and it is only relatively recently (in the past few millennia), that we have been able to modify our environment to any significant extent. Scientists and historians are sufficiently impressed by our 21st century environment modifying abilities to name the geological age we are in the "Anthropocene"[49].

The interplay between our environment and ourselves is mirrored in us by the interplay between our exteriors and interiors—our bodies and our minds. From an individual perspective we are able to track three key developmental stages. In the first two stages, pre-personal and personal, we are operating out of the reptilian brain and the limbic system respectively.[1] The R-complex and limbic systems effectively correspond to System 0 (the core bodily functions), and System 1 (the intuitive, analog intelligence which has often been called "right-brained").

---

[49] Many scientists are now using the term and the Geological Society of America titled its 2011 annual meeting: Archean to Anthropocene: The past is the key to the future. The Anthropocene has no precise start date, but based on atmospheric evidence may be considered to start with the Industrial Revolution (late 18th century). Other scientists link it to earlier events, such as the rise of agriculture. Evidence of relative human impact such as the growing human influence on land use, ecosystems, biodiversity and species extinction is controversial, some scientists believe the human impact has significantly changed (or halted) the growth of biodiversity

The bridge from the personal to the transpersonal phases of development is the neocortex, which in man constitutes two-thirds of our brain. A person without a neocortex is essentially a "vegetable" as they kindly put it in the medical business, while a mouse without a neocortex can function normally for all intensive purposes, so the development of the functions of the neocortex is crucial for humans to evolve into responsible, conscious, choice-making citizens in the highly complex world we have evolved over the past few thousand years.

We find System 2, the digital, procedural intelligence, emerging in this collaboration between System 0 and System 1, enabling us to program our actions and behaviours much more effectively, in the pursuit of specific goals, and in collaboration with others.

In the transpersonal phase of development the pre-frontal and frontal cortices become new control centres which enable individuals to activate System 3, the "strategic psychological helicopter", from which they are capable of getting an overview of the situation they are in and in which they are capable of being mindful and taking better decisions that benefit the greatest span of human beings while honouring the greatest depth in those individuals.

1. **Pre-personal Phase**—in our first decade or so of life we develop through the pre-personal phase. In the developed world this stage results in the emergence of a self that is distinct from our family, and in the developing world a self that is distinct from our tribe. Our heroes during this stage of development are powerful, often impulsive, egocentric and, of course, heroic. During the pre-personal phase we believe in Santa Claus, magical-mythic spirits, dragons, beasts, and powerful people. In ancient times these were archetypal gods and goddesses, while in the 21st century we find super-heroes and their archenemies embodied in books and film, from Harry Potter to Lord of the Rings to Batman. While we are developing our rational faculties during this phase, unconscious motivations and magical thinking often characterize our behaviour.
Dysfunction during these years can lead to issues arising throughout a lifetime. Challenging or primitive life conditions can also make it difficult for people to move beyond these developmental stages, and during times of war, famine, great hardship or personal difficulty,

people often regress to these stages as mature adults. In such cases the family or the tribe is the ultimate shock absorber.
2. **Personal Phase**—At some point between being a child and becoming a teenager, we learn to be conscientious, responsible people, if all goes well, and experience a desire to conform to conventional norms and behaviours. With the emergence of our own conscious identity and rational mental faculties, we are now able to strike a balance between our intuitive (System 1) and reasoning thought (System 2) processes at a conscious level. We may even be experiencing some System 3 moments where we transcend ourselves and the situations we are in, gaining a helicopter perspective and perhaps even a "peak experience" with spiritual Aha! Moments.

Modern educational systems, modern organizations and institutions and some religions are powerful forces in helping to shape such conventional and rational mental processes. Those who have been shaped by such systems often reach the achievement and Affiliative levels by their mid-twenties and form the majority of the population in most developed societies. Modern "civil" society is based upon conscious mental processes being activated and used on a daily basis by most people.
3. **Transpersonal Phase**—In the transpersonal realm one transcends and integrates all other developmental levels, moving through the authentic integral level to the transcendent and unity levels of higher consciousness. According to recent research, hundreds of millions of people are now actively exploring these transpersonal levels worldwide, while at least 1% of the world's population is anchored at the transpersonal in their daily life.

This is the great leap psychologist Clare Graves was talking about in 1970 as he reviewed his latest research results with Abraham Maslow and the rest of the American Psychological Association members. Such post-modern stages of development are the basis upon which an integral, global civilization could be built in the 21st century. I say could because:

- the billions of conformist and achievement oriented power holders would have to be sufficiently attracted to the possibilities that they are able to let go of some of their narrow belief systems and vested interests to give the newer systems and structures room to grow;

- the hundreds of millions of affiliative, cultural creatives would need to become much more grounded and practical in their desire for transformation and demands for change, while also shedding the last remnants of their often narcissistic tendencies.

Diagram 3 below offers a simplified representation of the different pathways the evolution of human consciousness can take. This model simplifies the three main, overlapping models of human development currently used in a wide variety of organizations, psychology and philosophy textbooks. The names of the successive stages follow the classification system used by psychologist Jenny Wade and the colors used by Professor Don Beck, the co-creator of Spiral Dynamics.

The "Reactive" (or "beige") stage evolved about 100 000 years ago when our distinct sense of self began to emerge, and food, water, warmth, sex and safety were our top priorities. We used our very wide range of senses (far wider than the five main senses used today), instincts and habits to survive in the wilderness, and formed into small survival bands to perpetuate life, living off the land close to nature along with the other animals. The original Bushmen of the Kalahari Desert, aboriginal tribes on various continents including Australia and the remaining tribes of the rainforests in the Amazon, Borneo and elsewhere living in small bands of up around 30 people are typical of this stage of development, where the most highly developed part of the brain is the R-complex.

Around 50 000 years ago, a new stage of development began to emerge when we started to invent art, music and religion—in short, the beginnings of culture and metaphorical thinking that transformed our ability to communicate and coordinate ourselves in larger groups ranging up to 300 people. This new tribal form of existence placed a priority on the individual being a loyal group member, showing allegiance to the tribal chiefs, elders, ancestors and the clan.

The Trouble with Paradise

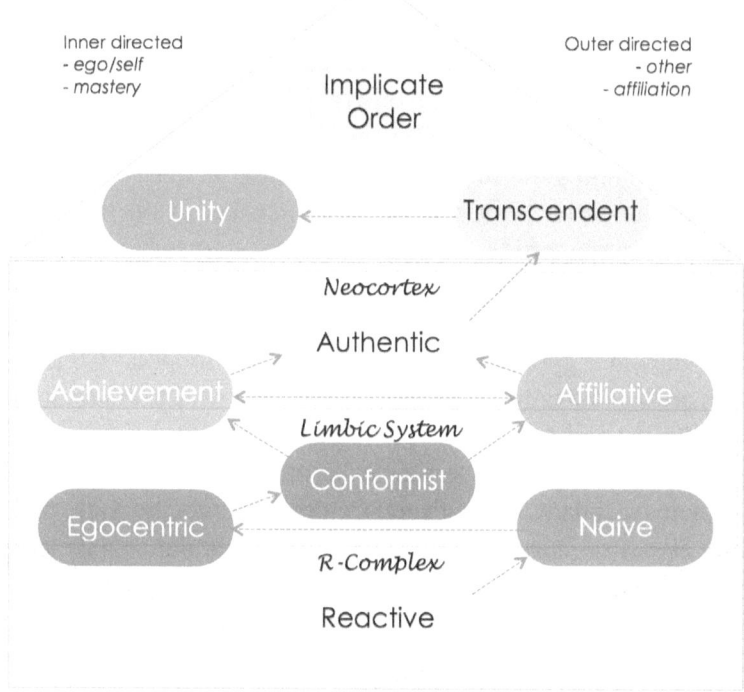

Diagram 3-9 Stages in the Evolution of Human Consciousness

At this naïve (or "purple") stage of development, sacred objects, places, events and memories are paramount to making sense of the world. Observing the rites of passage, seasonal cycles and tribal customs are essential for tribal cohesion and bonding between its members. Spirit beings and mystical signs are paid close attention to, ensuring that individuals have a sense of belonging and safety in the face of an unpredictable and often cruel world. People at this stage of development are outer-directed towards others, and value their affiliation with the group above all else. The 400 or so original tribes of Europe and the 400 or so original tribes of Africa were all interconnected through a variety of familial and trading alliances, yet also engaged in warlike behaviour frequently when life conditions got tough or over-crowded.

About 10 000 years ago, the egocentric (or "red") stage of development began to emerge. Like the magical/naïve purple stage of development, this stage is a halfway house between the R-complex part of our brains and the limbic system, which processes emotions and stores a treasure house of memories. Unlike the magical/naïve purple stage, however, this egocentric red stage is driven by a focus on self and the mastery of power over others. Through our development of a variety of technologies 10 000 years ago, the world became at once a more settled and yet a more dangerous place as massive armies and empires began to be built for the first time, based on the surpluses generated through settled urban populations harnessing a variety of agricultural, building and military technologies, as well as elaborate social hierarchies.

The ancient Egyptians, Greeks, Persians, Mongols, Chinese and Romans are good examples of Empire builders who came to dominate parts of Europe and Asia for several hundred years at a time, only to be pushed back to their native territories as other more powerful Empires rose in their ascendant. In South and Central America the Aztec and Inca Empires also rose more recently and exhibited similar social structures and power dynamics, though with very different life conditions and cultural content.

Egocentric, "Red" stands tall, expects attention, demands respect and calls the shots, and attempts to dominate competitors by conquering or out-foxing them. Being what you are and having what you want, regardless of the consequences for others, is the primary goal of such individuals and cultures. The emphasis of life is to enjoy oneself to the fullest right now without guilt or remorse, and breaking free from any domination or constraints as required to fulfil immediate desires. Egocentric red is thus highly impulsive, generally opportunistic, and often unpredictable.

About 5 000 years ago, major religions began to arise and spread around the planet, beginning with Egyptian and Chinese religions. During the Axial Age which began around 3 000 BCE and ended around 700 CE, an intense burst of new religion creation began starting with Buddhist and Hindu traditions, followed by Judaism, Christianity and Islam. Over the course of these 6 000 years, a new way of being emerged centered around a variety of belief systems encapsulated in "new stories", that gave life a meaning, direction and purpose with pre-determined outcomes. It must be stressed that not only are these religious belief systems—they also represent the emergence of new capacities in the human brain and being to recognize and participate in new forms of order that also have consequences for governance systems and culture, along with technological and scientific advances appropriate to that level of development.

Each religion has its founding leaders and/or gods, from Buddha and his three universal truths to the three principal gods of Hinduism—Shiva, Vishnu and Shakti, Moses and his God Yahweh, followed by the Hebraic Kings and Prophets and Jesus and Mohammed, both believed to be sons of God and god-like by Christians and Muslims respectively.

Such Conformist "True Blue" belief systems require a degree of sacrifice to their specific transcendent Cause, Truth or Righteous Pathway. There is a strong hierarchical order in which the powerful enforce a code of conduct based on eternal, absolute principles that, if followed and obeyed, produce stability now and guarantee future reward, whether in heaven or on earth. The impulsivity of the egocentric red way of being is now controlled through guilt and a desire to "do the right thing", where there is a place for everyone, and everyone is in their place (though not necessarily wildly happy about it). Laws, regulations and discipline build character and moral fibre, such that social and economic arrangements become more certain, predictable and enforceable.

The conformist stage of development is the last stage centred predominantly in the limbic system. Depending upon an individual's predisposition, the next stage of development can either be achievement or affiliation oriented. Some individuals might be balanced evenly between these developmental stages, though women tend to develop more strongly in the direction of affiliation, and men tend to develop more strongly in the direction of achievement.

The shift from the limbic system to the neocortex as the control centre that determines our priorities and creates our options and actions is a major advancement in human civilization. This was precisely the shift that occurred during the first European Renaissance starting in about 1500 in northern Italy. It has taken fully 500 years for the "Achiever" and "Affiliator" modes of consciousness to mature into fully-fledged ways of operating that form the centre of gravity of our modern and post-modern cultures in the developed world.

The Achievist ("Orange") level of development began to put down deep roots in European circles about 300 years ago. At its root this kind of consciousness is optimistic, risk-taking and stresses science, technology and competitiveness. It acknowledges that change and progress are built into the nature of things, and focuses on learning the secrets of nature and implementing the best solutions to hitherto difficult challenges. This modernist approach manipulates the earth's resources to create and spread abundance, without recognizing limits to growth. Achievist strategies are built around acting in one's own self-interest by playing the game to win—an approach Adam Smith would have applauded and which he set out clearly in his book "The Wealth of Nations" a few centuries ago.

It is no surprise, therefore, to learn that the Affiliative level of development exists in a strong tension with the Achievist level of development. This communitarian/egalitarian ("Green") mode of consciousness emerged slightly later than and often in reaction to aspects of the achievist "orange" system. It can be seen in the better parts of the French Revolution, the American constitution, the early Quakers, the movement for the abolition of slavery, and various European philosophers of the eighteenth and nineteenth centuries. Where the achievist level seeks to create wealth, the affiliative system is concerned with the equitable distribution of wealth, and is associated with Socialist and other liberal movements over the centuries.

The focus of green affiliators is on seeking peace within and without, co-creating caring in communities where feelings and sensitivity supersede cold rationality. One of the major contributions of "green" consciousness and movements it that it redresses the imbalances created by harsh blue TruthForce systems and often too greedy orange Achievist systems, helping heal old wounds and bringing reconciliation, truth and justice to the fore. The affiliative green system was at the forefront of the South African peace and reconciliation process after the harsh years of apartheid, enabling Nelson Mandela to be the "good guy" who took care of all South Africans, not just his own supporters.

The six levels of development from the Reactive Beige system through Magical Tribal Purple, Egocentric Red, Conformist Blue, Achievist Orange and Affiliative Green, are what is known as "First-Tier" systems. This is because they all share the common characteristic of believing that they are each "the only game in town". As one ascends from Beige through to Green, the higher levels systems take their own superiority for granted, as they are more complex and extend further in space and time than each of the previous levels. Blue TruthForce believes that if it could only evangelise everyone else effectively, then the "others" would sign up to their belief system or culture. Orange Achievists believe that everyone secretly wants progress above all else, and that people will do anything to get ahead. Green Affiliators believe that orange achievists are greedy and overbearing, and need to see the error of their ways, while those Blue Conformists are just so stuck in the mud and backward.

In reality, however, each value system is an adaptation to a specific niche, and in its own ways will be better or worse adapted depending upon specific life conditions prevailing in specific niches at certain points in time. The extent to which there is "dissonance" between people's values and their life conditions will determine to what extent their dissatisfaction with what no longer works for them will translate into action that leads them to evolve a more appropriate blend of values for their needs.

With the completion of the "Green Meme", human consciousness is poised for a quantum jump into "Second-Tier thinking." Clare Graves referred to this as a "momentous leap," where "a chasm of unbelievable depth of meaning is crossed." In essence, with second-tier consciousness, one can think both vertically and horizontally, using both hierarchies and heterarchies (both ranking and linking). One can therefore, for the first time, vividly grasp the entire spectrum of development, and thus see that each level is critical for the health of the overall Spiral. The Authentic, Integral, level of development is the starting point for this exciting journey, followed by the Transcendent and Unity stages of development. The latter two stages are not well mapped yet, as there are so few people in them at the moment.

Each wave of development transcends its predecessor, and yet it includes or embraces it in its own makeup. For example, a cell transcends but includes molecules, which transcend but include atoms. To say that a molecule goes beyond an atom is not to say that molecules hate atoms, but that they love them: they embrace them in their own makeup; they include them, they don't marginalize them. Each wave of existence is a fundamental ingredient of all subsequent waves, and thus each is to be cherished and embraced.

Moreover, each wave can itself be activated or reactivated as life circumstances warrant. In emergency situations, we can activate red power drives; in response to chaos, we might need to activate blue order;

in looking for a new job, we might need orange achievement drives; in marriage and with friends, close green bonding. All of these value systems have something important to contribute.

But what none of the first-tier value systems can do, on their own, is fully appreciate the existence of the other value systems. Each of the first-tier value systems thinks that its worldview is the correct or best perspective. It reacts negatively if challenged. Blue order is very uncomfortable with both red impulsiveness and orange individualism. Orange individualism thinks blue order is for suckers and green egalitarianism is weak and woo-woo. Green egalitarianism cannot easily abide excellence and value rankings, big pictures, hierarchies, or anything that appears authoritarian and thus green reacts strongly to blue, orange, and anything post-green.

All of that begins to change with second-tier thinking. Because second-tier consciousness is fully aware of the interior stages of development—even if it cannot articulate them in a technical fashion—it steps back and grasps the big picture, and thus second-tier thinking appreciates the necessary role that all of the various value systems play. Second-tier awareness thinks in terms of the overall spiral of existence, and not merely in the terms of any one level.

Where the green value system begins to grasp the numerous different systems and pluralistic contexts that exist in different cultures (which is why it is indeed the sensitive self, i.e., sensitive to the marginalization of others), second-tier thinking goes one step further. It looks for the rich contexts that link and join these pluralistic systems, and thus it takes these separate systems and begins to embrace, include, and integrate them into holistic spirals and integral meshworks. Second-tier thinking, in other words, is instrumental in moving from relativism to holism, or from pluralism to integralism.

The Integral Stage of Development—MindShift and WorldShift

The European philosopher Jean Gebser and American philosopher Ken Wilber have written a great deal about what they call the integral stage of development—Wilber to the extent that he calls his life's work "Integral Philosophy". If Wilber and others such as Paul Ray, researcher and author of the "Cultural Creatives" are right, then roughly one hundred million plus individuals are engaged in a transition from first-tier ways of being and thinking, into second tier or "Integral" ways of being and thinking. What would that mean for the evolutionary trajectory of our species Homo sapiens, and what would that mean for ThriveAbility?

Wilber's Integral Theory places perspectives at the centre of his way of describing our world and Kosmos. Psychologist Don Beck has also explored this developmental level using colourful terminology for the first stages of integral development (which he calls "Yellow", and then "Turquoise" for the more holistic, planetary perspective). Wilber makes some grand claims for his integral approach—here is the introductory page on Amazon for his book "Integral Vision":

"Suppose we took everything that all the various world cultures have to tell us about human potential—about psychological, spiritual, and social growth—and identified the basic patterns that connect these pieces of knowledge. What if we attempted to create an all-inclusive map that touches the most important factors from all of the world's great traditions?

Ken Wilber's Integral Vision provides such a map. Using all the known systems and models of human growth—from the ancient sages to the latest breakthroughs in cognitive science—it distills their major components into five simple elements, and, moreover, ones that readers can verify in their own experience right now.

In any field of interest, such as business, law, science, psychology, health, art, or everyday living and learning—the Integral Vision ensures that we are utilizing the full range of resources for the situation, leading to a greater likelihood of success and fulfilment. With easily understood explanations, exercises, and familiar examples, The Integral Vision shows how we can accelerate growth and development to higher, wider, deeper ways of being, embodied in self, shared in community, and connected to the planet, which can literally help with everything from spiritual enlightenment to business success to personal relationships."

We do indeed live in an extraordinary age, where all of the world's knowledge and cultures, are available to each of us—something unprecedented in the history of our species. For the past few hundred thousand years a person was born into a culture that knew only of its own existence. For example, someone born in Africa, was raised as an African, married an African, and followed an African religion—often living in the same hut for their entire life, on a spot of land that their ancestors settled millennia ago. We have evolved out of Africa from isolated hunter gatherer bands like the Bushmen, to the tribal settlements and farms of the Xhosa, to the ancient nations of Africa, to the conquering feudal empires of the Zulus and other warring tribes. In the past five hundred years we have also witnessed the emergence of international corporate states such as the Dutch East India Company to modern multinationals such as Unilever, and the other inhabitants of the global village including the United Nations and hundreds of thousands of NGO's and charities. We are now witnessing the next stage in our evolution toward an integral global village that seems to be humanity's destiny.

As Wilber puts it: "So it is that the leading edge of consciousness evolution stands today on the brink of an integral millennium—or at least the possibility of an integral millennium, where the sum total of extant human knowledge, wisdom, and technology is available to all. But there are several obstacles to that integral embrace, even in the most developed populations. Moreover, there is the more typical or average mode of consciousness, which is far from integral anything, and is in desperate need of its own tending. Both of those pressing issues (the integral vision as it relates to the most developed and the least developed populations) are related directly to the contents of this volume of the Collected Works."

Lifestyles And Creations

The *lifestyles and creations* we bring into the world all depend on the carrying capacity of the planet for their aliveness and sustainability. Today humanity is already exceeding planetary boundaries, so we need to reinvent our way of living and accelerate sustainable innovation to create breakthroughs everywhere.

Our lifestyles and creations:

- *Levels of Development*—are shaped by the level of development we and those close to us are centered in;
- *Human Capital*—are enabled or limited by the education, training and resources available to develop our own talents and human capital;

- *Lines of Development*—emerge from our strengths and capabilities in different lines of development[50];
- *Life Conditions*—are a function of the life conditions we have available to us, in particular the time, energy and resources needed to invest in ourselves and our creations, as well as our capacities to meet and transcend the many challenges life presents us with;
- *Institutional Support*—will thrive and grow based on the institutional systems and support available to us in the sector/s we operate in;
- *Footprint*—will have an environmental and social footprint related to both the industry/sector we are in and the extent to which we are conscious of and have access to renewable energies, resilient habitats and sustainable approaches applicable to our work and sector.

There is a great deal more that could be said about the topic of human and economic development implicit in "LifeStyle and Creations" ingredient of the ThriveAbility dashboard, but in the interests of brevity we will move on to consider the broader implications of the environmental and social footprint mentioned in the last point above about Footprint.

Carrying Capacity

Carrying capacity stands for the total sum of resources and flows available to support life in abundance, without compromising the Earth's ability to regenerate and recover from damages. This is the fundamental boundary condition for all actions. For life to survive and thrive, the carrying capacity has to be greater than our lifestyle and creations.

The most fundamental aspects of carrying capacity are measured by the Stockholm Resilience Centre, in its Planetary Boundaries research.

---

[50] What psychologist Howard Gardner called "Multiple Intelligences", which are generally physical, emotional, mental and/or spiritual in nature. Careers Guidance Counselors and testing systems differentiate between thousands of different kinds of skills that require combinations of these intelligences or lines of development plus some very specific competencies and tools.

In 2009, a group of 28 internationally renowned scientists identified and quantified a set of nine planetary boundaries within which humanity can continue to develop and thrive for generations to come. Crossing these boundaries could generate abrupt or irreversible environmental changes. Respecting the boundaries reduces the risks to human society of crossing these thresholds. The nine planetary boundaries are, briefly:

- **Stratospheric ozone layer**—The stratospheric ozone layer in the atmosphere filters out ultraviolet (UV) radiation from the sun. If this layer decreases, increasing amounts of ultraviolet radiation will reach ground level. This can cause a higher incidence of skin cancer in humans as well as damage to terrestrial and marine biological systems. Fortunately, because of the actions taken as a result of the Montreal Protocol, we appear to be on the path that will allow us to stay within this boundary.
- **Biodiversity**—The Millennium Ecosystem Assessment of 2005 concluded that changes in biodiversity due to human activities were more rapid in the past 50 years than at any time in human history, increasing the risks of abrupt and irreversible changes to ecosystems. The drivers of change that cause this severe biodiversity loss and lead to changes in ecosystem services are either steady, showing no evidence of declining over time, or are increasing in intensity.
- **Chemicals dispersion**—Emissions of toxic compounds such as heavy metals, synthetic organic pollutants and radioactive materials, represent some of the key human-driven changes to the planetary environment. These compounds can persist in the environment for a very long time, and their effects are potentially irreversible. Even when the uptake and bio-accumulation of chemical pollution is at sub-lethal levels for organisms, the effects of reduced fertility and the potential of permanent genetic damage can have severe effects on ecosystems. For example, persistent organic compounds have caused dramatic reductions in bird populations and impaired reproduction and development in marine mammals.
- **Climate Change**—Recent evidence suggests that the Earth, now passing 400 ppmv CO_2 in the atmosphere, has already transgressed the planetary boundary and is approaching several Earth system thresholds. We have reached a point at which the loss of summer polar sea-ice is almost certainly irreversible. This is one example

of a well-defined threshold above which rapid physical feedback mechanisms can drive the Earth system into a much warmer state with sea levels metres higher than present. The weakening or reversal of terrestrial carbon sinks, for example through the ongoing destruction of the world's rainforests, is another potential tipping point, where climate-carbon cycle feedbacks accelerate Earth's warming and intensify the climate impacts. A major question is how long we can remain over this boundary before large, irreversible changes become unavoidable.

- **Ocean acidification**—Around a quarter of the CO2 humanity emits into the atmosphere is ultimately dissolved in the oceans. Here it forms carbonic acid, altering ocean chemistry and decreasing the pH of the surface water. This increased acidity reduces the amount of available carbonate ions, an essential 'building block' used by many marine species for shell and skeleton formation. Beyond a threshold concentration, this rising acidity makes it hard for organisms such as corals and some shellfish and plankton species to grow and survive. Losses of these species would change the structure and dynamics of ocean ecosystems and could potentially lead to drastic reductions in fish stocks.
- **Freshwater consumption and the global hydrological cycle**—The freshwater cycle is strongly affected by climate change and its boundary is closely linked to the climate boundary, yet human pressure is now the dominant driving force determining the functioning and distribution of global freshwater systems. The consequences of human modification of water bodies include both global-scale river flow changes and shifts in vapour flows arising from land use change. These shifts in the hydrological system can be abrupt and irreversible. Water is becoming increasingly scarce—by 2050 about half a billion people are likely to be subject to water-stress, increasing the pressure to intervene in water systems.
- **Land system change**—Land is converted to human use all over the planet. Forests, wetlands and other vegetation types have primarily been converted to agricultural land. This land-use change is one driving force behind the serious reductions in biodiversity, and it has impacts on water flows and on the biogeochemical cycling of carbon, nitrogen and phosphorus and other important elements. While each incident of land cover change occurs on a local scale,

the aggregated impacts can have consequences for Earth system processes on a global scale. A major challenge with setting a land use boundary is that it needs to reflect not just the absolute quantity of unconverted and converted land but also its function, quality and spatial distribution.

- **Nitrogen and phosphorus inputs to the biosphere and oceans**—The biogeochemical cycles of nitrogen and phosphorus have been radically changed by humans as a result of many industrial and agricultural processes. Nitrogen and phosphorus are both essential elements for plant growth, so fertilizer production and application is the main concern. Human activities now convert more atmospheric nitrogen into reactive forms than all of the Earth's terrestrial processes combined. Much of this new reactive nitrogen is emitted to the atmosphere in various forms rather than taken up by plants. When it is rained out, it pollutes waterways and coastal zones or accumulates in the terrestrial biosphere. Similarly, a relatively small proportion of phosphorus fertilizers applied to food production systems is taken up by plants; much of the phosphorus mobilized by humans also ends up in aquatic systems. These can become oxygen-starved as bacteria consume the blooms of algae that grow in response to the high nutrient supply. A significant fraction of the applied nitrogen and phosphorus makes its way to the sea, and can push marine and aquatic systems across ecological thresholds of their own. One regional-scale example of this effect is the decline in the shrimp catch in the Gulf of Mexico's 'dead zone' caused by fertilizer transported in rivers from the US Midwest.

- **Atmospheric aerosol loading**—An atmospheric aerosol planetary boundary was proposed primarily because of the influence of aerosols on Earth's climate system. Through their interaction with water vapour, aerosols play a critically important role in the hydrological cycle affecting cloud formation and global-scale and regional patterns of atmospheric circulation, such as the monsoon systems in tropical regions. They also have a direct effect on climate, by changing how much solar radiation is reflected or absorbed in the atmosphere. Humans change the aerosol loading by emitting atmospheric pollution (many pollutant gases condense into droplets and particles), and also through land-use change that increases the release of dust and smoke into the air. Shifts in climate regimes and monsoon systems have already been seen in highly polluted environments. A further reason for an aerosol boundary is that aerosols have adverse effects on many living organisms. Inhaling highly polluted air causes roughly 800,000 people to die prematurely each year.

The WBCSD's Vision 2050 program has also defined a set of pathways to ensure we reduce our current "overshoot" of environmental and social boundaries, based on a specific set of industry analyses.

The result of extensive dialogues involving 200 companies spanning 20 countries, Vision 2050 has at its core the attributes of successful business planning: understand your current situation, identify the obstacles to success, and create a pathway to overcome those obstacles. The conclusion of this analysis is the need for a fundamental transformation of the way the world produces and consumes everything from energy to agricultural products. And in that shift, Vision 2050 identifies unprecedented opportunities for business—at least those that understand they can no longer operate in business-as-usual, autopilot mode.

Opportunities range from developing and maintaining low-carbon, zero-waste cities, to improving and managing biocapacity, ecosystems, lifestyles and livelihoods. In today's dollars, the market opportunities created by adapting to the new global reality for sustainable living are somewhere between $3-$10 trillion USD per year in 2050.

Vision 2050 is not only about economics, development and sustainability challenges for business. It suggests governments and civil society must create a different view of the future, one where, "economic growth has been decoupled from ecosystem destruction and material consumption and re-coupled with sustainable economic development and societal well-being."

With 9 billion people on the planet competing for a limited supply of natural resources, the definition of "living well" will also have to shift. Instead of a utopian dream, living well in 2050 means that all people have access to and the ability to afford education, healthcare, mobility, the basics of food, water, energy and shelter, and consumer goods. It also means living within the limits of the planet itself.

Sometimes the simplest questions are the hardest to answer. Vision 2050 asks those questions and offers a way to help businesses understand the pathways they will need to succeed. The question of where we will be in 2050 is well worth asking, for the rewards to those who get the answers right is unprecedented.

If Vision 2050 was to be successfully implemented by the WBCSD members and their business ecosystem partners and supply chains, we would reduce our current need for 1.5 planets of resources to 1.1 planets, which might also help us to:

- Ensure global warming does not exceed 2 degrees centigrade by helping . . .
- Reduce annual greenhouse gas emissions below 10 gigatonnes of CO_2 equivalent p.a. through shifts on both supply and demand sides
- Reduce waterstress, peak soil and overfishing effects while reducing air and water pollution & deforestation by incentivising sustainable supply chains & lifestyles
- Enhance the quality of life in and renewability of urban habitats, in particular in the world's 600 largest & fast growing cities
- Encourage entrepreneurs and community programs with positive social impacts in these cities through empowering, for-impact entrepreneurial activities

Predictive Modeling of Transformation to a Conscious Economy

Ingredient four is *predictive modeling of transformation to a conscious economy*. This refers to new methods and tools that are developed through an integration of different models and frameworks to provide hands-on support to put the ThriveAbility ideas into practice. These different models and tools are illustrated below in relation to the ThriveAbility Model, and provide the basis for the ThriveAbility Index ("GTI" in the diagram).

Diagram 4: GTI- Activating ThriveAbiity

A Final Word

ThriveAbility is a massive topic which also means we run the risk of grand visions which are initially very exciting, but ultimately never realised. One the of the commitments I am personally going to make as the Founder of ThriveAbility and co-founder of the ThriveAbility Consortium to anyone who gets involved in this initiative, is that we are going to be laser-focused in the implementation of the key apps and applications of ThriveAbility to ensure we do get traction and deliver effective outcomes.

Our initial focus will be on the ThriveAbility Index, which is to be prototyped during 2014 as one of the first projects within the newly established ThriveAbility Foundation. Much planning is now going into taking the first steps to make this a reality—if you have any questions or suggestions please e-mail me at rlw777@me.com, and I will do my best to get back to you subject to my workload and lifeload.

Finally, whatever errors are in this document are definitely my own responsibility, so please let me know if there is anything amiss. The views contained in this document are my own personal perspective, though as you will note from the credits, a number of very talented people have made some very important contributions, including our ThriveAbility Consortium members. Thanks to Paul van Schaik for his help with a number of the diagrams herein, which have been beautifully transformed from their initial powerpoint versions, and to Ken Wilber for his inputs.

Thanks for taking the time to read this far—I do hope you will find some way, no matter how small, to apply some of the thinking and models developed so far. Thrive Away!

CREDITS

My own ThriveAbility journey would not have been possible without encouragement from and contributions by many people. What follows is a short list of those who have been more closely involved than most, including the participants of the Embedding ThriveAbility programs of 2013 and 2014.

The ThriveAbility Consortium

Ralph Thurm—A|head Ahead	Previously head of Sustainability at Siemens, Deloitte & COO of the GRI
Christopher Cooke, Sheila Cooke	5 Deep Integral Ltd— Founders—OnlinePeopleScan

Embedding ThriveAbility Participants

Nicholas Beecroft	Psychiatrist, UK Ministry of Defense, Author—Future of Western Civilization
Laura Bechthold	MSc Student in Sustainability
Graham Boyd	LTS Global Managing Director
John Elkington	Founder SustainAbility, Chairman of Volans, "B-Team" Board Member
Pauline Engelberts	Global Head of Investment Products at ABN Amro
Chris Laszlo	Associate Professor at Case Western Reserve University
Ervin Laszlo	Co-Founder—Club of Budapest and Club of Rome
Sabine Oberhuber	Co-founder at Turntoo Foundation (circular economy)
Tim Odell	Founder—GiveAll2Charity
Lars Schipholt	Integrality BV
Sebastian Straube	Co-Founder and CEO—BSD Consulting, Germany

Dr Robin Lincoln Wood

Tiia Tammaru	Chairman of the Board of Estonian Quality Association
Elena Wood	Co-Founder—Chateau La Tour Apollinaire and Galatea SARL
Greg Wood	Co-Founder and Sales Director—Boardex

And a special word of thanks to two supporters from the very beginning:

Tani Jarvinen	Founder and CEO—Lautukeskus Excellence Finland
Kari Keskinen	Vice-President—Lautukeskus Excellence Finland

And to:

Paul van Schaik	Co-Founder of Integral Without Borders and IntegralMentors
Ken Wilber	Integral Philosopher and Founder—Integral Institute

ENDNOTES

[i] Production of the elements from iron to uranium occurs within seconds of a supernova explosion. The metal from items such as rings and bracelets were synthesized at this time, billions of years ago in the history of the universe. Due to the large amounts of energy released in a supernova explosion, much higher temperatures are reached than stellar temperatures. These higher temperatures allow for an environment where transuranium elements might be formed. In nuclear fusion processes in stellar nucleosynthesis, the maximum weight for an element fused is that of iron, reaching an isotope with an atomic mass of 56. Fusion of elements between silicon and iron occurs only in the largest of stars, which end as supernova explosions (see Silicon burning process). A neutron capture process known as the s process which also occurs during stellar nucleosynthesis can create elements up to bismuth with an atomic mass of approximately 209. However, the s process occurs primarily in low-mass stars that evolve more slowly.

[ii] Pat Ament. Climbing Everest, McGraw Hill 2001.

[iii] Genesis 1.1: In the beginning God created the heavens and the earth. Now the earth was formless and empty, darkness was over the surface of the deep, and the Spirit of God was hovering over the waters. And God said, "Let there be light," and there was light. God saw that the light was good, and he separated the light from the darkness. God called the light "day," and the darkness he called "night." And there was evening, and there was morning—the first day.

[iv] Douglas Adams, author of the Hitchhikers Guide to the Galaxy, was apparently influenced by John Cleese in selecting "42" and his daughter Polly (born when Douglas Adams was, you guessed it, 42).

[v] Bohm, David, Wholeness and the Implicate Order, London: Routledge, 1980

[vi] Michael Talbot. The Holographic Universe, HarperCollins (1991)

[vii] Carried out by the global polling organization IPSOS Mori and the Tony Blair Faith Foundation in 2010.

[viii] While 30% of the world's population adhere to other religions, and 16% are non-religious.

[ix] As discovered by Rocky Raccoon in his hotel room, in the song of that name by the Beatles in the White Album, 1967. "Gideon had left it there no doubt, to help in young Rocky's revival".

x   The best-selling novel of the *21st century*, with nearly 100 million copies sold at the date of writing this book.

xi   Linguistic evidence suggests the Romanies in Northern Europe originated from the Rajasthan people, emigrating from India towards the northwest no earlier than the 11th century. Contemporary populations sometimes suggested as sharing a close relationship to the Romani are the Dom people of Central Asia and the Banjara of India. Who exactly the Perpignan Gitanes are related to is anyone's guess, though there is definitely an Indian connection.

xii   Gruber and Kersten (1995) claim that Buddhism had a substantial influence on the life and teachings of Jesus. They claim that Jesus was influenced by the teachings and practices of Therapeutae, described by the authors as teachers of the Buddhist Theravada School then living in Judaea. They assert that Jesus lived the life of a Buddhist and taught Buddhist ideals to his disciples; their work follows in the footsteps of the Oxford New Testament scholar Barnett Hillman Streeter, who established as early as the 1930s that the moral teaching of the Buddha has four remarkable resemblances to the Sermon on the Mount."

Some scholars believe that Jesus may have been inspired by the Buddhist religion and that the Gospel of Thomas and many Nag Hammadi texts reflect this possible influence. Books such as The Gnostic Gospels and Beyond Belief: the Secret Gospel of Thomas by Elaine Pagels and The Original Jesus by Gruber and Kersten discuss these theories.

xiii   Officially endorsed by the Catholic Church around 1129, the Order became a favored charity throughout Christendom, and grew rapidly in membership and power. Templar knights, in their distinctive white mantles with a red cross, were among the most skilled fighting units of the Crusades. Non-combatant members of the Order managed a large economic infrastructure throughout Christendom, innovating financial techniques that were an early form of banking, and building many fortifications across Europe and the Holy Land.

The Templars' existence was tied closely to the Crusades; when the Holy Land was lost, support for the Order faded. Rumors about the Templars' secret initiation ceremony created mistrust, and King Philippe IV of France, took advantage of the situation. Under pressure from King Philippe, Pope Clement V disbanded the Order in 1312. The abrupt disappearance of a major part of the European infrastructure gave rise to speculation and legends, which have kept the "Templar" name alive into the modern day

xiv   The Temple Church is a late 12th century church in London located between Fleet Street and the River Thames, built for and by the Knights Templar as their English headquarters. In modern times, two Inns of Court (Inner Temple and

The Trouble with Paradise

Middle Temple) both use the church. It is famous for its effigy tombs and for being round. The area around the Temple Church is known as the Temple and nearby is Temple Bar and Temple tube station.

[xv] Chateau La Tour Apollinaire in Perpignan—see www.latourapollinaire.com for more information and history. The Chateau is today a luxury bed and breakfast, and the chapel a beautiful bedroom fit for a Post-Modern Queen.

[xvi] In Catholic theology, an indulgence is the full or partial remission of temporal punishment due for sins which have already been forgiven. The indulgence is granted after the sinner has confessed and received absolution. They are granted for specific good works and prayers. Indulgences replaced the severe penances of the early Church. Alleged abuses in selling and granting indulgences were a major point of contention when Martin Luther initiated the Protestant Reformation (1517).

[xvii] Including copies distributed free by Church groups and Gideon as part of their evangelical mission.

[xviii] Précis of The Archives Of The Cape Of Good Hope-1715-1806. The Library of The University Of California Los Angeles "Borcherds (Meent) ; Minister of Stellenbosch ; has found that difficulties are attached to transporting to him his allowances, and therefore requests to be given Rds. 20 instead, the amount granted some years ago to his colleagues at the Cape because of the excessive dearness of house rent and provisions. He believes that his request will be considered reasonable, as his colleagues here have certainly no more arduous duties to perform than he has, who receives a most unequal share of remuneration for it, they receiving in cash Rds. 60 each per month, and he only Rds. 10. It is true that formerly Rds. 20 were allowed for house rent, but they were afterwards withdrawn, when 1790. The Minister had been provided with a parsonage, but for all that he receives Rds. 30 less than the others. It may be that you suppose that living is much cheaper in Stellenbosch than in the Capital, and also that much profit is derived from the vineyard adjoining the Parsonage, and therefore may deem his re-quest less reasonable than it really is.

He therefore remarks, regarding the first point, how the special conditions of this village, so different from those of all the three others require an expenditure, equivalent to that at the Cape, for he leaves it to your consideration, whether all other things left aside, he is not to pay as much for his corn, meat, clothes, and other necessaries, at Stellenbosch as at the Cape, yea! Even much more for some things. And as regards the profits of the vineyard, he acknowledges that if there was water leading to the Parsonage as before, it would to some extent pay him, but it is at present in such a state, that the profits, after deducting the expenses for Cellar

Material, Slaves, etc., so necessary for its maintenance, are annually of not much account. He therefore prays that, instead of the emoluments hitherto received by him from the Pantry, he may be given Rds. 20 per month, and that among them may also be reckoned those of the last half year. He further notifies that he has now concluded his five years' contract, having left Europe on the 23rd December, 1784, and wishing to renew it, asks for the usual increase."

[xix] Nevertheless, his dealings with the missionaries in Stellenbosch, Mewes Jans Bakker and Erasmus Smit, were full of conflict. The problems were ecclesiological. Borcherds considered it his duty to protect the rights of the Dutch Reformed Church, its parish council and ministers. He believed that the minister ultimately was responsible for guaranteeing the orthodoxy of religious services held under his auspices, so he had to examine and, in effect, license the missionaries in his parish. The services which they held should not, of course, conflict with those of the established church. Furthermore, he argued that it was only the minister of the parish church who had the right to baptize, therefore those who had been prepared for baptism by Bakker had to be passed on to him for the final examination and the administration of the ceremony.

The friction this standpoint caused with the equally principled, if personally less forceful and socially secure, Bakker, can easily be imagined. Essentially these were matters of ecclesiastical law, so it was decisions of government which ultimately determined the relationship. In the first instance, the dictates of Commissioner-General J.A. Uitenhage de Mist's church ordinance were heavily in Borcherds' favour. As a representative of the rationalist Batavian government, De Mist saw religious enthusiasm as a threat to the social and political order. Therefore the church ordinance re-established the virtual monopoly of the Dutch Reformed Church and forbade the extension of missions, except under the auspices of its ministers. In this way De Mist hoped to maintain government control over religious affairs.

This was not a line which the British colonial rulers could accept after their reconquest of the Cape in 1806. Eventually, ordained missionaries were given the right to act independently and to baptize their converts themselves, a decision which Borcherds finally accepted with good grace. The result of the conflict, though, was to leave the missionaries very much in the position of junior partners in the ecclesiastical ranking of the Western Cape towns.

[xx] "Historic Houses of South Africa" By Dorothea Fairbridge (with a Preface by General J. C. Smuts), Oxford University Press, 1922—The original of this book is in the Cornell University Fine Arts Library—NA7468.6.S6F16 ISBN 3 1924 014 905 834

[xxi] General Smuts continued: Such are usually found in what are called the older countries. It will, therefore, probably come as a surprise to the reader of this book, who is not a South African, to find houses and estates dating back to within a century of van Riebeeck having the appearance of a mellowed antiquity. This book is an attempt to preserve or, at any rate, to record what is most noteworthy in our older South African architecture and domestic surroundings.

Those who have seen the awful destruction of the Great War and the absolute obliteration of everything in what were some of the most beautiful districts of Europe will appreciate the necessity for recording by pen and pencil the works of a period in South Africa while these remain to us. Even in this uncrowded land the hand of the builder and restorer is heavy, and even while I write, some beautiful building may be defaced."

This is often done simply because the attention of the would-be improver has not been directed to the beauties of what he possesses, and he does not see that what is consecrated by the taste of one age is not lightly to be touched by the hand of another. The old houses of South Africa are a common heritage of which all South Africans are proud, and are precious links binding us all together in noble traditions and great memories of our past.

From the tragedy which has convulsed the older world we look with thankfulness at our own South Africa, with her mysterious compelling attraction, her peace, the great gifts that Providence has showered upon her. The youngest of the sister nations which form the British Empire, she may take her place with dignity amongst them, sorrowfully proud in her sons who have died to uphold her good name and maintain her honour and fealty."

[xxii] From the accounts of Pieter in his autobiography.

[xxiii] Auto-Biographical Memoir—Civil Commissioner Of Cape Division And Resident Magistrate For Cape Town And District Thereof, And Cape District. A Plain Narrative Of Occurrences From Early Life To Advanced Age, Chiefly Intended For His Children And Descendants, Countrymen And Friends.

Cape Town: A. S. Robertson, Adderley-Street. 1861. Dedicated By Permission To Sir George Grey, K.C.B.,

Governor And Commander-In-Chief At The Cape Of Good Hope, High Commissioner Of British Kaffraria,

By His Most Humble and Obedient Servant, P. B. Borcherds, Senior.

Dr Robin Lincoln Wood

Preface.

[xxiv] In the November 1997 issue of *Natural History*, (p. 19).

[xxv] Woody Allen's character, Sid Waterman, in the filmmaker's latest comic-suspense movie, *Scoop*.

[xxvi] Marshall McLuhan, The Global village—page 94.

[xxvii] To understand entrepreneurship and leadership, Dr Dave Robinson has integrated Beck and Cowan (Graves' theories) with ethics and organizational psychology models to create the Personal and Corporate Values Journey 'PCVJ' diagram (1998). After several decades of research, Robinson goes further in his phenomenological approach to tie it to several other cultural, business, and logical paradigms (mainly within entrepreneurial business environs) and suggests leadership tools for communication and growth of subordinates and self, linking heavily to Gravesian interpretations. The PCVJ was first presented academically at the 2007 AGSE conference

[xxviii] In cell biology, a mitochondrion is a membrane-enclosed organelle found in most eukaryotic cells. Mitochondria are sometimes described as "cellular power plants" because they generate most of the cell's supply of adenosine triphosphate (ATP), used as a source of chemical energy. In addition to supplying cellular energy, mitochondria are involved in a range of other processes, such as signaling, cellular differentiation, cell death, as well as the control of the cell cycle and cell growth. Mitochondria have been implicated in several human diseases, including mitochondrial disorders and cardiac dysfunction, and may play a role in the aging process

[xxix] Of interest to the genealogist (among others) is the fact that all of an individual's mitochondria are derived from his/her mother. Although the sperm cell tail is packed with mitochondria to power its long journey to the egg cell, the tail and mitochondria drop off of the sperm at fertilization and never enter the egg cell. Consequently, all of the mitochondria in the fertilized egg come from an individual's mother.

[xxx] What we humans call "food".

[xxxi] In each cell of your body there are one or more organelles called *mitochondria* whose job is to take your fats and sugars and turn them into energy. This energy is needed by the cell to do its job properly and to provide energy for the full body. When the mitochondria are working properly, you will have all the energy you need, all of your organs and cells will be working appropriately, and unless you are consuming many more calories than you need, you should be very close to your optimal weight.

Unfortunately, almost all of the chemical and heavy metal toxins that are in each of us are potent mitochondrial toxins. These common toxins include: mercury, arsenic, cadmium, lead, chlorinated pesticides, organophosphate pesticides, PCBs, dioxins, aromatic hydrocarbons (from combustion) solvents and plasticizers. Mitochondrial dysfunction not only leads to generalized fatigue, but is now documented to be a major factor in the development of many of the most serious chronic illnesses in our time (including cancer, neurological diseases like Parkinsonism, diabetes, etc.).

[xxxii] The composition of our atmosphere remains fairly constant providing the ideal conditions for contemporary life. All the atmospheric gases other than noble gases present in the atmosphere are either made by organisms or processed by them. The Gaia theory states that the Earth's atmospheric composition is kept at a dynamically steady state by the presence of life. The stability of the atmosphere in Earth is not a consequence of chemical equilibrium as in in planets without life. Oxygen is the second most reactive element after fluorine, and should combine with gases and minerals of the Earth's atmosphere and crust. Traces of methane (at an amount of 100,000 tonnes produced per annum) should theoretically not exist, as methane is combustible in an oxygen atmosphere.

Dry air in the atmosphere of Earth contains roughly (by volume) 78.09% nitrogen, 20.95% oxygen, 0.93% argon, 0.039% carbon dioxide, and small amounts of other gases including methane. While air content and atmospheric pressure varies at different layers, air suitable for the survival of terrestrial plants and terrestrial animals is currently known only to be found in Earth's troposphere and artificial atmospheres.

[xxxiii] Processing of the greenhouse gas CO2 plays a critical role in maintaining the Earth's temperature within the limits of habitability. The CLAW hypothesis, inspired by the Gaia theory, proposes a feedback loop that operates between ocean ecosystems and the Earth's climate. The hypothesis proposes that the particular phytoplankton that produce dimethyl sulfide are responsive to variations in climate forcing, and that these responses lead to a negative feedback loop that acts to stabilize the temperature of the Earth's atmosphere.

Currently this Gaian homeostatic balance is being pushed by our rising of human population and the impact of our activities on our environment. The rise of greenhouse gases appears to be turning Gaia's negative feedbacks into homeostatic positive feedback. According to Lovelock, this could bring an accelerated global warming and mass human mortality.

[xxxiv] "The Vanishing Face of Gaia", Basic Books, 2009.

[xxxv] If we can get CO2 to stay below 450 ppm in the next 25 years we may have a chance of staving off the worst effects of runaway climate change: the worst case

scenario is that we have a planet capable of sustaining 1 billion people by 2100. Even when one life is lost through negligence that is sad—but a few billion lives in less than a century would be unspeakable. And that is the trajectory we are heading toward with business and life as usual.

We have already wiped out one-quarter of all species on earth, and the faster the temperature rises, the faster the world's largest mass extinction will happen, and it will include most of us beyond a 4 deg C warming scenario. In the past 60 years our generation has left a legacy which may prove deadlier than all the wars in the 20th century. The unsustainable growth of our economies during this period has driven most of our planet's ecosystems to the point of collapse:

- Two-thirds of our global fishing grounds are over-exploited
- Half our coral reefs have died
- 35% of the world's population living on the 41% of the earth's surface which are dry lands are now running out of water
- One-quarter of all species have become extinct, and half of the remaining species face extinction this century
- We've pumped enough CO_2 into the atmosphere to ensure a minimum global warming of two degrees centigrade, and are on course to warm the planet by a total of four degrees by 2050.

Right now we are on course for at least a 3-4 deg C warming scenario. Even without runaway ice cap melting and methane release "doom loops", we are in for one hell of a ride. We are already experiencing rising ocean levels, massive deforestation, habitat destruction, increased disease transmission, mass migrations, reduced water availability, increased natural hazards especially weather-related disasters, and catastrophic changes in ocean chemistry and ability to sustain life as we know it. That's the result of 1 deg C warming and our current bad habit of using up every square inch of productive biosphere on the planet and pumping almost all of the hundreds of millions of years of accumulated carbon deposits out of the ground and pumping them back into the air in less than two centuries. There is, however, somewhere between denial and fear, an opportunity: to use this crisis to create a second Renaissance on a global scale.

Now for the good news. The combination of peak oil, the emergence of the digital knowledge economy and integral culture and leadership mean that we now have global awareness among decision-makers about the challenges we face, even if they do not agree in detail about the optimal strategy for addressing them. And we have tens of millions of leaders thinking and acting globally, rather than simply trying to make a short-term profit at the planet's expense. For the sake of the planet we must empower these leaders by connecting them up with each other and with the new generation of leaders who are being asked to demonstrate a heroism and wisdom beyond their years.

xxxvi   In 1750, before the industrial revolution, atmospheric CO2 was stable at 280 ppm. By 1960 it had reached 315 ppm, and is currently (in 2011), around 394 ppm. The earth has warmed by about a degree centigrade since 1750, and even if we kept CO2 at current levels (which is nigh on impossible), the delayed ocean warming effect of all that CO2 will be another 0.5 deg C.

Despite the best efforts of a small but determined group of people over the past three decades, we are in a situation where the Kyoto 1 protocol, together with more sustainable policies at national and local levels and major efforts by 20% of large businesses, are failing to halt the inexorable rise in greenhouse gas emissions which threaten all of us. Before the end of the century it is possible that between 300 million to one billion people could die as a result of climate change, and billions more will suffer. In Africa alone it is estimated that 184 million people will succumb, while in Asia-Pacific estimates are in the range of the hundreds of millions. And that is simply based on global temperature increases already in the pipeline from the greenhouse gases we have already produced to date, plus our current annual greenhouse gas output doing business and life as usual. See my book: "The Great Shift—Catalyzing the Second Renaissance" for more information: http://amzn.to/rlwgreatshift

xxxvii   Thanks to "Intel Inside", probably one of the best known information technology advertising campaigns ever.

xxxviii   Initially sulfur aerosols and volcanic ash enveloped the earth's atmosphere, blocking out sunlight and sending surface temperatures plunging. Ash and sulphur aerosols can remain in the upper atmosphere for hundreds to thousands of years, which would be enough to cause a significant glaciation. At the end of the Permian period the biggest ever drop in sea level in history occurred. Two scientists (named Holser and Magaritz) proposed in 1987 that such a marine regression could be caused by a large scale glaciation.

xxxix   In the Paleocene epoch—as if you are going to remember that.

xl   The "punctuated equilibrium" theory of Niles Eldredge and Stephen Jay Gould was proposed as a criticism of the traditional Darwinian theory of evolution. Eldredge and Gould observed that evolution tends to happen in fits and starts, sometimes moving very fast, sometimes moving very slowly or not at all. On the other hand, typical variations tend to be small. Therefore, Darwin saw evolution as a slow, continuous process, without sudden jumps. However, if you study the fossils of organisms found in subsequent geological layers, you will see long intervals in which nothing changed ("equilibrium"), "punctuated" by short, revolutionary transitions, in which species became extinct and replaced by wholly new forms.

[xli] The Alps arose as a result of the collision of the African and European tectonic plates. Enormous stress was exerted on sediments of the Alpine Tethys basin and its Mesozoic and early Cenozoic strata were pushed against the stable Eurasian landmass by the northward-moving African landmass. Technically, the Alps extend all the way to the Himalayas.

[xlii] The Precambrian spans from the formation of Earth around 4600 million years ago to the beginning of the Cambrian Period, about 542 million years ago, when macroscopic hard-shelled animals first appeared in abundance.

[xliii] Ben Tipping, *Exemplary Epic: Silius Italicus' Punica* (Oxford University Press, 2010), pp. 20-21

[xliv] For example, in Esperaza in the Aude, there is a unique dinosaur museum displaying some of the spectacular finds made nearby

[xlv] Known as 'La Caune de l'Arago'.

[xlvi] In our post-modern world the cranial capacity required to operate computers, perform delicate surgery, compose symphonies, pilot aircraft and write detective novels exceeds 1,400 cubic centimeters.

[xlvii] **Punctuated equilibrium** is more an observation than a theory of evolution. However, this observation is easy to explain by using some general insights from the systems approach. Consider a typical fitness landscape, in which there are valleys separated by ridges. If the evolving system has reached the bottom of a deep valley, there will be almost no change, since variation will fail to pull the system out of that hole. This is a negative feedback regime, in which chance fluctuations will be counteracted, pulling the system back to its equilibrium position at the bottom of the valley.

On the other hand, if there is only a small ridge separating the valley from a neighboring, deeper valley, then a chance event may be sufficient to push the system over the edge so that it enters the other valley. Such a lucky variation will become increasingly likely when the fitness landscape changes so as to reduce the height of the ridge. Once over the ridge, the descent into the new valley will go very fast. This is a positive feedback regime in which deviations from the previous position are amplified. This means that the system will evolve very quickly to a new, fitter configuration. If we would check the evolution of the species in the geological record, we would find many fossils corresponding to the position at the bottom of the valley where the organism remained for so long, but few or none corresponding to the crossing of the ridge, which happened very fast on the geological time scale.

The systems approach can help us to understand more profoundly how a small variation can produce a major change. Indeed, organisms, like all systems, are organized in levels, corresponding to their subsystems and sub subsystems. Each subsystem is described by its own set of genes. A mutation in one of the components at the lower levels will in general have little effect on the whole. On the other hand, a mutation at the highest level, where the overall arrangement of the organism is determined, may have a spectacular impact. For example, a single mutation may turn a four-legged animal into a six-legged one. Such high-level mutations are unlikely to be selected, but potentially they can lead to revolutionary changes.

A fundamental example of such a major change is the metasystem transition, where a system evolves in a relatively short time to a higher level of complexity.

[xlviii] According to evolutionary systems scientist Ervin Laszlo, "[culture] evolved from expressive signs, such as animals use to communicate, to denotative symbols, typical of human languages. Whereas signs provide a stimulus which signals something of immediate significance in the communicator's environment, a symbol may have a meaning which is entirely divorced from the here-and-now." The symbols Laszlo refers to are the sounds made by the mouth, the pictures on the cave wall, or the images made by the hand in sign language—these are mostly external artifacts, not yet inter-subjective understandings.

[xlix] The neurologist Paul MacLean has proposed that our skull holds not one brain, but three, each representing a distinct evolutionary stratum that has formed upon the older layer before it, like an archaeological site. He calls it the "triune brain." MacLean, now the director of the Laboratory of Brain Evolution and Behaviour in Poolesville, Maryland, says that three brains operate like "three interconnected biological computers, [each] with its own special intelligence, its own subjectivity, its own sense of time and space and its own memory". He refers to these three brains as the neocortex or neo-mammalian brain, the limbic or paleo-mammalian system, and the reptilian brain, the brainstem and cerebellum. Each of the three brains is connected by nerves to the other two, but each seems to operate as its own brain system with distinct capacities.

This hypothesis has become a very influential paradigm, which has forced a rethink of how the brain functions. It had previously been assumed that the highest level of the brain, the neocortex, dominates the other, lower levels. MacLean has shown that this is not the case, and that the physically lower limbic system, which rules emotions, can hijack the higher mental functions when it needs to.

It is interesting that many esoteric spiritual traditions taught the same idea of three planes of consciousness and even three different brains. Gurdjieff for example referred to Man as a "three-brained being". There was one brain for the spirit, one for the soul, and one for the body. Similar ideas can be found in Kabbalah, in Platonism, and elsewhere, with the association spirit—head (the actual brain), soul—heart, and body in the belly. Here we enter also upon the chakra paradigm—the idea that points along the body or the spine correspond to nodes of consciousness, related in an ascending manner, from gross to subtle.

The Reptilian Brain. The archipallium or primitive (reptilian) brain, or "Basal Brian", called by MacLean the "R-complex", includes the brain stem and the cerebellum, is the oldest brain. It consists of the structures of the brain stem—medulla, pons, cerebellum, mesencephalon, the oldest basal nuclei—the globus pallidus and the olfactory bulbs. In animals such as reptiles, the brain stem and cerebellum dominate. For this reason it is commonly referred to as the "reptilian brain". It has the same type of archaic behavioural programs as snakes and lizards. It is rigid, obsessive, compulsive, ritualistic and paranoid, it is "filled with ancestral memories". It keeps repeating the same behaviours over and over again, never learning from past mistakes (corresponding to what Sri Aurobindo calls the mechanical Mind). This brain controls muscles, balance and autonomic functions, such as breathing and heartbeat. This part of the brain is active, even in deep sleep.

The Limbic System (Paleomammalian brain). In 1952 MacLean first coined the name "limbic system" for the middle part of the brain. It can also be termed the paleopallium or intermediate (old mammalian) brain. It corresponds to the brain of the earliest mammals. The old mammalian brain residing in the limbic system is concerned with emotions and instincts, feeding, fighting, fleeing, and sexual behaviour. As MacLean observes, everything in this emotional system is either "agreeable or disagreeable". Survival depends on avoidance of pain and repetition of pleasure.

When this part of the brain is stimulated with a mild electrical current various emotions (fear, joy, rage, pleasure and pain etc) are produced. No emotion has been found to reside in one place for very long. But the Limbic system as a whole appears to be the primary seat of emotion, attention, and affective (emotion-charged) memories. Physiologically, it includes the hypothalamus, hippocampus, and amygdala. It helps determine valence (e.g., whether you feel positive or negative toward something, in Buddhism referred to as vedena—"feeling") and salience (e.g., what gets your attention); unpredictability, and creative behaviour. It has vast interconnections with the neocortex, so that brain functions are not either purely limbic or purely cortical but a mixture of both.

MacLean claims to have found in the Limbic system a physical basis for the dogmatic and paranoid tendency, the biological basis for the tendency of thinking to be subordinate feeling, to rationalize desires. He sees a great danger in all this limbic system power. As he understands it, this lowly mammalian brain of the limbic system tends to be the seat of our value judgements, instead of the more advanced neocortex. It decides whether our higher brain has a "good" idea or not, whether it feels true and right.

The Neocortex, cerebrum, the cortex, or an alternative term, neopallium, also known as the superior or rational (neomammalian) brain, comprises almost the whole of the hemispheres (made up of a more recent type of cortex, called neocortex) and some subcortical neuronal groups. It corresponds to the brain of the primate mammals and, consequently, the human species. The higher cognitive functions which distinguish Man from the animals are in the cortex. MacLean refers to the cortex as "the mother of invention and father of abstract thought". In Man the neocortex takes up two thirds of the total brain mass. Although all animals also have a neocortex, it is relatively small, with few or no folds (indicating surface area and complexity and development). A mouse without a cortex can act in fairly normal way (at least to superficial appearance), whereas a human without a cortex is a vegetable.

The cortex is divided into left and right hemispheres, the famous left and right brain. The left half of the cortex controls the right side of the body and the right side of the brain the left side of the body. Also, the right brain is more spatial, abstract, musical and artistic, while the left brain more linear, rational, and verbal.

[i]   http://en.wikipedia.org/wiki/Ervin_L%C3%A1szl%C3%B3

Science and the Akashic Field: An Integral Theory of Everything—Ervin Laszlo, (Inner Traditions, 2007. 2nd edition)

The Chaos Point: The World at the Crossroads (Hampton Roads, 2006); Quantum Shift in the Global Brain: How the New Scientific Reality Can Change Us and our World [Rochester VT: Inner Traditions, 2008]; WorldShift 2012: Making Green Business New Politics & Higher Consciousness Work Together (McArthur & Company, 2009)

[ii]   The Case of Steve Jobs—The "War of the Worldviews" is alive and well between conventional medicine, represented here by the priesthood of "Star Surgeons", and medically trained people recommending alternative treatments that may have a much better success rate, but lack the credibility of the mainstream. (For example, chemotherapy is ineffective 97% of the time, yet still the most profitable product for doctors and pharmaceutical companies). Medicine is a huge, high growth business which is growing rapidly globally, along with the western,

materialistic worldview and consumerist values that drive our pill-popping, quick fix culture. That can often seem like an unstoppable force to those who have to make life and death choices.

Ironically and tragically, Steve Jobs' instinctive response to his initial diagnosis was to try an alternative treatment when he should have had immediate surgery to prevent the cancer spreading, Then, it would have been highly appropriate to use alternative treatments to heal and prevent a recurrence. This is a serious problem for patients, who have to choose (like Steve Jobs did, incorrectly and sadly as it turns out) between competing claims in their treatment. Seriously ill people should not have to bear that burden, let alone well people seeking preventive input and treatments.

The War if the Worldviews is a tragedy at every conceivable level, for millions of people everywhere, every day. If mindbody medicine and proven alternative treatments had been integrated in a single healthcare delivery system, Steve Jobs might still be alive and well today, playing with his children and changing the world.

[lii] Dr Lipton has published several books including: 2006 The Wisdom of Your Cells—How Your Beliefs Control Your Biology and 2009 Spontaneous Evolution: Our Positive Future and a Way to Get There from Here

[liii] http://en.wikipedia.org/wiki/Bruce_Lipton

[liv] http://edgemagazine.net/2008/02/norm-shealy/

[lv] http://www.amazon.com/War-Worldviews-Where-Science-Spirituality/dp/0307886891

[lvi] http://www.deanradin.com/NewWeb/EMindex.html

[lvii] http://www.amazon.com/Flourish-Visionary-Understanding-Happiness-Well-being/dp/1439190763

[lviii] Such as Jean Gebser—http://en.wikipedia.org/wiki/Jean_Gebser and Ken Wilber—http://en.wikipedia.org/wiki/Ken_Wilber

[lix] Such as Don Beck http://en.wikipedia.org/wiki/Don_Beck_(management_consultant), Robert Kegan http://en.wikipedia.org/wiki/Robert_Kegan & Jane Loevinger http://en.wikipedia.org/wiki/Jane_Loevinger

[lx] Such as Leonard and Murphy—http://en.wikipedia.org/wiki/Esalen_Institute. The mission of the Esalen Institute is: "Esalen Institute exists to promote the harmonious development of the whole person. It is a learning organization dedicated to continual exploration of the human potential, and resists religious,

scientific and other dogmas. It fosters theory, practice, research, and institution-building to facilitate personal and social transformation and, to that end, sponsors seminars for the general public; invitational conferences; research programs; residencies for artists, scholars, scientists, and religious teachers; work-study programs; and semi-autonomous projects."

[lxi] http://www.giordanobrunouniversity.com/ "The Giordano Bruno University is a humanistic online global institution, committed to create informed and ethical agents of change who will bring a new consciousness, a fresh voice and up-to-date thinking to the international community, transforming obsolete paradigms and empowering the co-creation of an equitable, responsible and sustainable world."

ABOUT DR ROBIN WOOD

Dr Robin Wood is an unusual blend of global thinker, entertainer, futurist, straight talking innovator and change agent - an explorer of the far frontiers of technology, business, culture and society. Robin takes his readers on journeys of discovery through the critical trends and possibilities that enable us to transform the future and realise our brilliance as individuals, organizations and societies. Robin brings his colorful adventures and insights gained from living and working in 35 countries on four continents to life with his dry sense of humour and sharp eye for detail. In his life and career he has worked with many business and world leaders driven by his passions for information, transformation and beneficial globalization. He is a frequent keynote speaker at conferences on a wide range of topics, including:

- Vision 2050: Co-Creating a Viable Future for a Smarter Planet
- The Future of Western Civilization and You
- Getting Your Act Together for the Great Shift
- Designing a Planet for a Second Renaissance

- The Evolution of Business 2.0—Creating a Living, Intelligent Economy
- Thrival-Beyond Sustainability and Resilience to ThriveAbility
- The Future of Strategy- Business Design and Business Ecosystems for the 21st Century
- How to Follow Your Bliss Without Losing Your Shirt—My Journey
- Leadership for the 21st Century
- FutureHealth- Transforming Healthcare

Robin is the author of several award-winning books and has been a Fellow at the Centre for Management Development at London Business School and at the Institute for Coherence and Emergence in the USA. Some of Robin's previous publications include:

The Great Shift- Catalyzing the Second Renaissance—*Renaissance2 Publications, 2010. (available from Amazon)*
Managing Complexity-How the New Sciences can Help Businesses Adapt and Prosper-*The Economist Books, 2000. (available from Amazon)*
Sunday Times *Book of the Week and* **Director Magazine** *Book of the Month*
Developing Leadership Capacity—Searching for the Integral—*Integral Leadership Review, 2003*
Seven Lessons from Microsoft—*Business Strategy Review, London Business School, 2003*

Dr Wood specializes in designing major shifts in socio-political and business ecosystems by inspiring, aligning and empowering the teams and organizations that make them happen. In the multi-trillion dollar Wintel personal and business computing ecosystem, for example, he worked with the senior executives in HP, Intel and Microsoft over a period of 5 years to design the futures of enterprise and personal computing, networking and storage, as well as reconfiguring HP's business design for the internet.

In the area of global socio-political systems, Dr Wood co-developed the World Bank's first Sustainable Energy policy, co-founded one of Europe's first e-business incubators in Ernst & Young, consulted to the UK Cabinet Office on social capital and the digital economy and contributed to the UN Year of Micro-credit. He also co-led the Strategy in Action programme for the top 100 executives in Vodafone over 3 years while a Fellow at the Center for Management Development at London Business School.

In the past two decades Robin has had the privilege to serve, advise and facilitate the leadership teams in governments and senior executives in more than 40 global organizations including the UK Ministry of Defence, The UK Cabinet Office, The S African Department of Health, The World Bank, 3M, Barclays, BBA Aviation, Citigroup, Eagle Star Insurance, Ernst and Young International, Hewlett Packard, HP/Microsoft alliance, HP/Intel alliance, ICL, Jardine Matheson, Royal & Sun Alliance, Kellogg's, Unilever, Royal Dutch PTT, Shell International, State Street Bank, Vodafone and the ZF Group.

Dr Wood spent over three years at Ernst & Young where he co-founded the strategy and e-business practices. During the heady days of the internet boom he became a Managing Director at e-business creator Scient. He has an ongoing relationship with business angels, incubators and leading social entrepreneurs, with special interests in artificial intelligence and knowledge management. During the first decade of his career Robin worked as a top corporate lawyer, Citicorp investment banker, Manager of Electronic Banking at TrustBank, Senior Consultant at PA Consulting in IT Strategy and Head of Marketing and Planning at the BIS Group (European HQ of NYNEX/Verizon).

His current focus as the Founder of the ThriveAbility Foundation is on embedding ThriveAbiity in organizations worldwide to ensure a bright and joyfully sustainable future for our emerging global civilisation. http://embeddingthriveability.org/

A keen amateur jazz pianist, Robin enjoys spending quality time with his wife Elena and his extended family including his two children Callum and Kirstie. He relaxes by walking in the Pyrenean mountains, swimming and snorkelling in the Mediterranean near Perpignan, France, where he's created a mini-oasis at Chateau La Tour Apollinaire. http://www.latourapollinaire.com/en/